Leafy Spices

Author
V. Prakash, Ph.D.
Scientist-in-Charge
Biophysical Chemistry Unit
Food Chemistry Department
Central Food Technological Research Institute
Mysore, India

CRC Press
Boca Raton Ann Arbor Boston

Library of Congress Cataloging-in-Publication Data

Prakash, V. (Vishweshwaraiah), 1951-
Leafy spices / author, V. Prakash.
p. cm.
Includes bibliographical references and index.
ISBN 0-8493-6723-9
1. Herbs. 2. Spice plants. 3. Aromatic plants. 4. Essences and essential oils. I. Title.
SB351.H5P73 1990
664′.53—dc20 90-40951
CIP

This book represents information obtained from authentic and highly regarded sources. Reprinted material is quoted with permission, and sources are indicated. A wide variety of references are listed. Every reasonable effort has been made to give reliable data and information, but the author and the publisher cannot assume responsibility for the validity of all materials or for the consequences of their use.

Direct all inquiries to CRC Press, Inc., 2000 Corporate Blvd., N.W., Boca Raton, Florida 33431.

International Standard Book Number 0-8493-6723-9

Library of Congress Card Number 90-40951
Printed in the United States

ACKNOWLEDGMENTS

I would like to express my deep sense of gratitude to Dr. B. L. Amla, Director, CFTRI, Mysore, India for his constant encouragement and support during the course of writing this book. My sincere thanks are due to Mr. C. P. Natarajan, former Director of CFTRI, Mysore for his helpful suggestions, critical comments, and supportive ideas during the entire preparation of this review. My thanks are also due to the staff of Spices and Flavor Technology Discipline of CFTRI, especially to Mr. N. B. Shankaracharya. My special thanks are also due to my wife Mrs. Jamuna Prakash for her patient and painstaking correction of proof and constant support during the entire preparation of this book.

PREFACE

New books on major spices from a number of authors are published frequently to cover chemistry, technology, and essential oils. However, the surge of literature is not massive on minor spices like ''leafy spices''; their importance nevertheless, is no less than many of the major spices. Up till now, a compact book on leafy spices was not available covering aspects such as botany, varieties, cultivars, harvesting, chemical composition, extraction of oil, and end uses. The present volume which has concise information on more than 20 leafy spices should cater to a large audience who are involved in active research in the area of spices to stimulate newer thinking and better utility for this class of spices.

V. Prakash
Mysore, India

THE AUTHOR

V. Prakash, Ph.D., is Scientist-in-Charge, Biophysical Chemistry Unit, Food Chemistry Department, Central Food Technological Research Institute, Mysore, India.

Dr. Prakash graduated in 1976 with a B.S., M.S., and Ph.D. degree from the University of Mysore, Mysore, India.

Dr. Prakash is a member of the Editorial Board of *Die Nahrung* journal, GDR, Association of Food Scientists and Technologists, India, Indian Biophysical Society, India and Indian National Committee for International Union of Pure and Applied Biophysics, and Guha Research Conference, India. Dr. Prakash has written a number of review articles in CRC publications and in *Methods in Enzymology*.

Dr. Prakash has given a number of extension lectures in several universities and institutions in India and abroad. During 1989-90 he was Visiting Professor (RA) in the Biochemistry Department, Brandeis University, Waltham, Massachusetts.

Dr. Prakash has received the Golden Jubilee Award for Science and Technology for his contribution in the area of food chemistry at the University of Mysore, during 1987. He is the recipient of the prestigious Sarma Memorial Award of the Society of Biological Chemists (India) for 1989.

Dr. Prakash has presented over 25 papers at international meetings and over 15 invited lectures at national meetings. He has published more than 70 papers. His current interest relates to drug-protein interactions, thermal stability of enzymes, and structure-property relationship of macromolecules with emphasis on functional properties of the food system and additives to it.

TABLE OF CONTENTS

Chapter 1

INTRODUCTION[1-52]

Spices and herbs have occupied an important place in the culinary preparations of several ancient and modern kitchens from time immemorial. One remembers India being known world over as the home of spices. Spices have been known for over 60 centuries or more. Most of the spices are valued not only for their flavor but also for their other special properties like stimulation of appetite and carminative action and preservation of food due to their special antimicrobial properties. Spices also play a significant role in the national economy in terms of export and import. The various areas in which the impact of a spice is commanding includes the food industry, medicine, pharmaceuticals, cosmetics, perfumery, and other related industries.

According to the International Organisation for Standardisation ''spices and condiments'' are defined as ''such natural plant or vegetable products or mixtures thereof in whole or ground form which are used for imparting flavour, aroma and piquancy to and for seasoning of foods''. On the other hand there is a wide discrepancy among experts over the definition of the simple word herb. Of several definitions, an herb can be stated as being ''a plant without woody tissue that withers and dies after flowering''; ''a plant or portion of a plant used in medicine, in cookery and for the extraction of essential oils''; ''a plant with little or no woody tissue without any winter buds which dies down to the ground at the end of the growing season''. Charlemagne humorously defines herb probably more precisely as: ''an herb is a friend of Physician and praise of Cooks''. Heath gives a botanical definition to the herb as ''soft stemmed plants whose main stem dies down to the root and regrows every year''.

With the above history of spices and the various definitions for herbs it may be right to point out some of the thoughts on the scientific basis of flavoring. It is well known that smell and taste are usually regarded as the most primitive animal senses (after touch); it is astonishing to note that we do not understand their scientific basis compared to that of sight and hearing of which we know a lot more. The need for a new beginning integrating the various aspects of chemistry, physics, mathematics, botany, and physiology emerging possibly to an area like ''sensory biophysics'' is of utmost importance. Such approaches would mean understanding of the taste of food especially related to spices and herbs from a more basic angle and quantification of the data even though the true tastes that are detected physiologically are very few, such as sourness, sweetness, saltiness, bitterness, astringency, and pungency. Our system can many times detect flavors arising from parts per million of specific ingredients in a food. This is what brings about the importance of minute addition of spices and herbs to a food during and after cooking to enhance its appeal and taste. Many times it is the use of the whole spice or the herb that brings about the characteristic flavor of a particular food. Obviously the same effect cannot be brought about by the usage of specific essential oils. The various chemical reactions that are brought about as a result of cooking, pickling, enzymatic reactions, oxidation, and interaction between the various ingredients in a food has a lot of bearing on the increase or decrease of flavor by usage of specific spices and herbs. I have been using the term ''herb'' very frequently to highlight its role in the world of flavor which generally gets missed among the major spices which predominate over many of the culinary preparations.

Since this review covers mainly the leafy spices, more emphasis is given to this in the next few pages so that the reader is gradually introduced to the various aspects of leafy spices.

There are several possible methods of classifying spices. They can be grouped according to (1) part of the plant from which the commercial products are produced, (2) their properties, and (3) families to which the plant belongs. Based on this broad classification, one comes across several subclassifications such as tropical spices, spicy seeds, herbs, aromatic vegetables, tree spices, major spices, leafy spices, pungent spices, phenolic spices, aromatic barks, and colored spices. Many methods of classification have been suggested by various authors but still no comprehensive, satisfactory, and strict classification has been possible based on agronomic, morphologic, or chemical grounds.

There are nearly 100 major spices grown in different parts of the world. If one considers strictly the botanical point of view of the various parts of the plant being used, one could classify the various spices as dried flower buds (cloves); fruits (allspice, black pepper, nutmeg, and vanilla); underground stem (ginger, horseradish, turmeric); bark (casia and cinnamon); seeds (anise, caraway, fenugreek, coriander, cardamom, dill, poppy, and cumin), and leaves (basil, bay, curry leaf, coriander leaf, mint, rosemary, sage, thyme, marjoram, oregano, savory, tarragon, and parsley).

Among the leafy spices, Heath further classifies them into five subgroups according to their sensory attributes i.e., those having a particular component as its prime constituent in its essential oil. They are group I cineole containing (sweet bay, laurel, rosemary, Spanish sage); group II thymol and/or carvacrol containing (thyme, wild thyme, oreganum, wild marjoram, sweet savory, and Mexican sage); group III containing sweet alcohols (sweet basil, sweet marjoram, tarragon); group IV containing thujone (Dalmation sage, Greek sage, and English sage), and group V containing menthol (peppermint, spearmint, garden mint, and corn mint). However, these do not include leafy spices such as curry leaf, coriander leaf, and other minor leafy spices.

There are nearly 200 plants officially listed as herbs and leafy spices. Of these however, only 20 to 30 may be considered commercially useful culinary herbs or spices. However, the U.S. Food and Drug Administration does not differentiate between culinary herbs and spices. It is all grouped together as spice.

The growing of herbs is very tricky and one has to have a good amount of experience since so much depends on soil and climate. Personally, I do grow a lot of leafy spices, of course permissible under South Indian climate/soil conditions and I feel it is an art by itself including the picking of the leaves at the proper time, just before flowering and maintaining the foliage.

From the production point of view the food industry is estimated to be consuming nearly 125,000 tons of spices with an annual worth about nearly £90 million (1990). Out of this there is no clear indication of how much is the contribution of leafy spices as such.

Many of the spices already have standards well laid out both for whole, ground, essential oils, and oleoresins. However, the need for uniform standards including microbiological specifications for various leafy spices is felt. These are being attempted at several national and international meetings. Now some specificiations are available in the U.S. and Canada for the dried leaves and essential oils of many of the leafy spices.

In this review the various leafy spices are dealt with in alphabetical order. They include basil, bay, bergamot, borage, burnet, camomile, chervil, chives, coriander, costmary, curry leaf, marjoram, mint, oregano, parsley, rosemary, sage, savory, tarragon, and thyme. By no means can the information be comprehensive but attempts are made to collect and represent facts in order, sometimes unpublished findings and observations also. The need for producing more of these leafy spices is very obvious and the increase in biomass leads to a higher yield of essential oil which is economically important. The information is cataloged into its geographical origin and presents status, botany, chemical composition, extractives, specifications if any, volatile oil and the chemistry of its constituents, specifications, technology, uses of the leafy spices in the food industry, medicine, pharmaceuticals, cosmetics, and perfumery. This it is hoped would help in understanding more about these precious group of spices which have a tremendous potential.

Chapter 2

BASIL

I. *OCIMUM BASILICUM* (L.) (SWEET BASIL) — FAMILY: LABIATAE

Basil, *Ocimum basilicum*, known also as sweet basil, is an annual spicy herb of the mint family and is indigenous to India. It is claimed that over 50 species are native to India. It is now produced commercially in France, Hungary, Indonesia, Morocco, and to some extent in almost all other warm and temperate countries in the world including the U.S. where California produces a superior quality of leaf. Hindus (a religious sect in India) plant it around their homes and temples to ensure happiness. In India, basil is little used in cooking but a type of basil (*Ocimum sanctum*) is regarded as holy tulsi by the Hindus.[1,3,4,12,52]

A. BOTANY[3,5,17,54,55,58,60-63]

It is an erect almost glabrous herb, 30 to 90 cm high and is a native of Central Asia and Northwest India. Leaves are ovate, lanceolate, cucuminate, toothed or entire, glabrous on both surfaces, and glandular. The flowers are white or pale-purple in simple or many-branched racemes that are often thyrsoid, the nutlets are ellipsoid black and pitted.

Mature leaves are ovate, reaching about 5 cm in length not including the petiole of about 2 cm. On the upper surface they are smooth and lustrous; on the lower surface along them midrib and on the petiole, short, stiff hairs occur sparingly.

The flowers are borne in long terminal racemose influorescences. The greenish corolla is small and inconspicuous. The calyx partly grown together with branches enlarges itself postflorally and remains with the latter dry on the plant. The capitate hairs have commonly a two-celled head with a stalk so short as to appear sesile. Chief structural characters of the leaves are ovate, pointed, blunt toothed when young, smooth above, and sparingly hairy below and along the mid rib.

Polymorphism and cross-pollination under cultivation have given rise to a number of subspecies and varieties differing in height, habitat and growth, degree of hairiness and color of stems, and in their leaves and flowers.[54]

B. CULTIVATION[53,54]

Ocimum basilicum is propagated by seeds and is commonly grown in gardens as an aromatic herb. The best season for sowing in the plains of India is October to November and in the hills from March to April. Seedlings are raised in the nursery beds and transplanted generally 30 cm apart and in rows spaced 40 cm apart. The crop is ready for harvest in 75 to 90 days after planting. Several cuttings may be made during the season. Singh et al.[54] recommend an optimum spacing of 40 × 60 cm and three cuttings at 2-month intervals beginning from 60 days after planting. This is supposed to give a higher yield than cutting at intervals of 3 to 5 months.

C. HARVESTING AND PROCESSING[53,54,58,68]

A yield of nearly 7 tons of leaves and flowers per hectare in two cuttings is reported from trial cultivations in Kanpur, India. Domestic yields are 8 to 10 tons of fresh herb per acre. The yield of green herb under favorable conditions can reach 12 tons per hectare. In Tarai, India, the optimum yield of herb is obtained by taking the first four crops of flowers only and the last crop of the entire flowering herb. The first harvest is taken when the plants are in full bloom, the second and subsequent harvests become available thereafter every 15 to 20 days. The last harvest is taken of the entire plant and oil distilled. While harvesting

TABLE 1
Composition of Sweet Basil (100 g, Edible Portion)[3,5,18]

Component	Quantity
Water	6.1—6.4 g
Food energy	251 kcal
Protein	11.9—14.4 g
Fat	3.6—4.0 g
Total carbohydrate	45.0—61.0 g
Fiber	17.0—18.0 g
Ash	14.3—16.7 g
Calcium	2113 mg
Iron	42 mg
Magnesium	422 mg
Phosphorus	490 mg
Potassium	3433 mg
Sodium	34 mg
Zinc	6 mg
Ascorbic acid	62 mg
Thiamin	0.1 mg
Riboflavin	0.3 mg
Niacin	6.9 mg
Vitamin A (as β-carotene)	9375 IU

the crop, precaution should be taken that the root-system of the plant is not injured, otherwise yield of subsequent harvest will be affected adversely.

When the leaves are to be marketed, they are dried artificially to preserve the green color at a temperature of less than 110°F under cover. The fresh leaves can be dried in the open air but more efficiently indoors by controlled artificial heat and circulating air. The dried leaves are reduced to specific-sized fragments by a series of machines and graded fine, medium, and coarse.

Dried sweet basil leaves have an aromatic, fragrant, sweet, and warm flavor and the taste is aromatic with a warm pungent undertone.

Drying in the sun and ordinary drying of basil do not affect much either the yield of essential oil or the content of phenols.

D. COMPOSITION

The composition of the dried leaf is shown in Table 1.[3,5,18]

E. SPECIFICATIONS IN THE U.S.[5]

Current specifications for spices do not include requirements for basil, either whole or ground.

A good commercial sample of sweet basil might be expected to contain volatile oil at a minimum of 0.4%; total ash (max.) 15%; acid insoluble ash (max.) 1.0%; moisture (max.) 8% and total ether extractives (min.) 4% on moisture free basis.

F. EXTRACTIVES[5]

The odor of essential oil is sweet, aniselike, cooling, and floral, whereas its nearest competitor, the exotic basil is harsh, penetrating, camphoraceous, phenolic, warm, and spicy with a lingering aftertaste.

The oleoresin of basil is prepared primarily from varieties having a high methyl chavicol content. The extract is very dark green, very viscous, almost solid, with a minimum volatile oil content of nearly 40 ml/100 g of the spice. The equivalency of the oleoresin to freshly ground basil in flavor and odor characteristics is 0.34 kg (0.75 lb) to 45.45 kg (100 lb).[5]

G. YIELD OF HERB OIL[3,5,6]

The flowers, on an average, yield 0.4% oil while the whole plant (Indian basil) contains 0.10 to 0.25% oil. By taking the initial three to four harvests of flowers (including main and subinfluorescence) and final harvests of whole herb one can get about 3 to 4 tons of flowers and about 13 tons of whole herb per hectare corresponding to about 13 kg of the so-called flower oil and about 27 kg of whole herb oil. In all, 40 kg of oil per hectare can be obtained. Oil of sweet basil, produced both from herb as well as flowers has a potential market.

H. DISTILLATION OF OIL[3,5,6]

Oil of sweet basil is produced by the distillation of the herb. The flowers or/and whole herbs are put into a distillation unit and hydro-distilled or steam-distilled. It takes about 4 h to complete one charge. The oil, being lighter than water, can easily be separated from an oil-water mixture. It is advisable to use the distillate after removing oil for further charge as it contains a small quantity of oil. Precaution should be taken to see that the distillation unit is clean and free from other odors, otherwise oil is likely to be contaminated with undesirable odors and colors.

I. VOLATILE OIL[56,62-65,72]

The yield of oil is nearly 2%. Sweet basil oil possesses a clovelike scent with an aromatic and somewhat saline taste. It yields a volatile oil (oil of basil) which is used as a flavoring agent and also as a perfume.

Two types of basil oil are commonly recognized.[56,63] One is the European type of sweet basil oil distilled in Europe and the U.S., containing linalool and methyl chavicol as the main constituents. It does not contain camphor. The oil exhibits levorotation. This type of oil has a fine odor and is considered superior in quality.

The Reunion type of basil oil produced in the islands of Camoros, Madagascar, and Seychelles contains methyl chavicol and camphor and does not contain linalool. The oil exhibits dextrorotation. This type of oil is considered inferior to the European type of sweet basil oil. It possesses a camphorous odor.

The other two minor types of basil oil are (1) the methyl cinnamate type — which is distilled in Bulgaria, Sicily, Egypt, India, and Haiti and which contains methyl chavicol, linalool, and a substantial amount of methyl cinnamate as its major constituents; and (2) the Eugenol type distilled in Java, Seychelles, and the U.S.S.R. which contains eugenol as the main constituent. The methyl cinnamate type is levorotatory and Eugenol type dextrorotatory.

Statistics on the production and utilization of basil oil in various countries are not readily available. According to available information, basil oil is not used in large quantities at the moment and the demand fluctuates. Perhaps its high market price indicates that the oil may be in short supply in many of the user countries.

J. COMPOSITION AND PROPERTIES OF VOLATILE OILS[3,56,58,59]

The various properties of the oils are presented in Table 2.

1. American Sweet Basil Oil (European type)[55]

Lowmann reported the following physicochemical properties of sweet basil oil produced at Arlington, Virginia. Yield: 31.5 lb of oil per acre; specific gravity at 15°: 0.9132 to 0.9278; refractive index 1.4883 to 1.4943; acid number 0.68 to 1.85; optical rotation $-6°21'$ to $-9°42'$; solubility 1 vol or more of 80% alcohol. Prakash[3] and Guenther[56] reported the following constituents in the American type of sweet basil oil: a terpene (b.p. 175 to 179°C and αD $+3°36'$) and small quantities of α-limonene, cineole, *l*-linalool; methyl chavicol, (30%) eugenol, and a sesquiterpene.

TABLE 2
Yield and Characteristics of Essential Oils of *Ocimum basilicum* from Different Growing Areas in India[53,54,56,63]

Properties	Delhi: Leaves and soft stems	Ghaziapur: Leaves and flowery tops	Bombay: Flowering plants	Kanpur: Leaves	Kanpur: Flowers	Jammu: Leaves	Jammu: Flowering tops
Yield %	0.5	0.4	0.35	0.16—0.34	0.31—0.54	0.53	0.44
Specific Gravity	0.9942 at 22°	0.9666 at 20°/20°	1.004—1.032 at 15°	0.9676—1.0040 at 24°	0.9718—1.0250 at 24°	0.9640	0.8700
n^{20}	1.5275	1.5122	1.3410—1.5405	1.513 to 1.527	1.5130 to 1.5280	1.5050	1.4625 at 30°
α_D	−11°	−4.9°	−4.2°—−3.0°	−5° to 6.8°	−4.6° to −6.6°	—	—
Acid value	11.1	2.0	1.1—1.2	0.6 to 2.2	0.3—5.8	0.7	0.2
Ester value	—	178.7	197.1—241.3	—	—	21.9	—
Ester value after acetylation	—	259.2	249.9—274.1	—	—	—	—
Ester (as methyl cinnamate) %	56.7	52	57.1—69.7	43.6—63.6	47.7—74.3	3.6 as linalyl acetate	—
Alcohol (as linalool) %	4.4	4.5	20.6—11.3	12.0—33.8	8.6—28.5	17.7	9.4

2. Reunion Type of Basil Oil[3,56]

The Reunion type of basil oil differs from true sweet basil oil by higher specific gravity, higher refractive index, and dextro- instead of levorotation. The yield of oil is about 30 kg/ha.

The chief constituents of the oil are methyl chavicol (34%) and *l*-linalool (55.4%). α- and β-pinene and 1,8-cineole are among the minor constituents.

The physicochemical properties of the oil are specific gravity at 15°/15°: 0.945 to 0.987; optical rotation −6°41′ to 0°46′; refractive index 20°: 1.5153 to 1.5182; saponification number 4.3 to 9.3; ester number after acetylation 28.9; acid number 0.2 to 3.0; total alcohol content 8.1%; solubility 3 to 7 vol of 80% alcohol. The oil also is shown to contain *d*-α-pinene, cineole, *d*-camphor, and methyl chavicol.[56]

3. Methyl Cinnamate Type of Oil[56]

The following values give the range for the above type of oil as reported by Guenther.[56] Specific gravity 15°/15°: 0.9309 to 0.9546; optical rotation −5°6′ to 7°6′; acid number 0.4 to 2.8; ester number (as methyl cinnamate) 40.1 to 126.9; ester number after acetylation 136.5; congealing point −3.9°C; solubility 1 vol of 80% alcohol; saponification number 46.1 to 61.83.

The chemical constituents of oil from both the small and large leaves:[54,55] cineole: 2.5%, *l*-linalool 40 to 48% methyl cinnamate 15 to 21.0%; eugenol 0.3 to 2.0%; methyl chavicol 21 to 23%; sesquiterpenes in small quantities.[55,56]

The leaves distilled along with stalks and flowering tops of *Ocimum basilicum* in Ghaziapur, India, gave a yield of 0.4% with properties as shown in Table 2.[55] The oil consisted of mainly *l*-linalool 45% (approx.) and methyl cinnamate: 52% (approx.). Nigam and Dutt distilled an oil of basil from the leaves of *Ocimum basilicum* and obtained a yield of 0.5%.[56] It's comprised of terpinene 20.85%; *l*-linalool 4.36%; methyl cinnamate, 56.67%. Analysis of a sample of Haiti basil (West Indies) indicated the following properties: specific gravity 0.966; optical rotation 2°5′; refractive index 20°: 1.5113; saponification number 93.0; ester content (methyl cinnamate) 27.0%; free alcohol content (linalool) 35.2%; solubility 1 vol of 80% alcohol.

The oil of *Ocimum basilicum* contains α-pinene, camphene, β-pinene, myrcene, limonene, 1,8-cineole, transocimene, 3-octanone, *p*-cymene, camphor, linalool, bornyl acetate, caryophyllene, terpinen-4-ol, methyl chavicol, α-terpineol, citronellol, eugenol, and methyl cinnamate.[86] They have also tentatively identified methyl eugenol in the same oil.[86]

Lawrence et al.[85] reported that the chemical composition and morphological characteristics of *O. basilicum* are very variable. The major constituents in the oil varied from one strain to another: *cis*-ocimene + 1,8-cineole trace 13.6%; linalool 0.2 to 75.4%; methyl chavicol 0.3 to 88.6%; methyl cinnamate trace 15.5%; eugenol trace 11.2%. Subsequently, Huang et al.[84] identified methyl chavicol, linalool, 1,8-cineole, ocimene — no isomer given, linalyl acetate, eugenol, *l*-epibicyclo-sesquiphellandrene, menthol, menthone, cyclohexanol, cyclohexanone, myrcenol, and nerol in an oil of basil of Chinese origin. The authors also showed that myrcenol and nerol possessed antiasthmatic activity.

4. Eugenol Type of Basil Oil[56,57]

Ocimum basilicum L. var. selasih mekah and selasih besar from Java upon distillation yielded oil 0.2 to 0.3% with the following properties: specific gravity at 26°: 0.890 to 9.940; optical rotation −11°15′ to 18°0′. The oil contained ocimene, but no myrcene and also 30 to 40% of eugenol. The analysis of the oil indicated the following ranges in their properties;[68] specific gravity 15/15° 0.9607 to 1.0027; optical rotation +1°57′ to −14°54′; acid number 4.7 to 10.2; ester number after acetylation 140.1; ester number 0 to 3.7; eugenol content 38 to 61%; phenol content 60.1%; solubility 7 vols of 90% alcohol or 0.7 vols of 80% alcohol.

5. French Sweet Basil Oil[56]

The producing regions of French sweet basil oil are Pegomas, Mandelieu, and Grasse. Production of the herb amounts annually to about 3,000,000 kg of which Pegomas itself produces half of this quantity.

Distillation (direct steam stills) of the leaves are generally done for 90 min. From 900 to 1100 kg of flower tops yields 1 kg of oil. The flowering whole overground parts of the plant, distilled at the end of the harvest contain considerably less oil, i.e., nearly 1400 to 1500 kg yielding about 1 kg of oil.

Methyl chavicol, the main constituent of sweet basil oil, is oxidized on aging and exposure of the oil to light and air, thus older oils usually show a higher specific gravity and higher refractive index. Sweet basil oil must, therefore, be stored carefully.

The curly leafed variety is said to give the best yield of oil and incidentally the finest quality.

6. Bulgarian Basil Oil[56,61]

The composition of Bulgarian oil was analyzed using preparative and sorption chromatography on aluminum trioxide and IR spectra. A preliminary separation of oil was made by fractional distillation. Gas-liquid chromatography of the oil was also used. The following compounds were found in the oil: *l*-linalool (main constituent); 1 to 8 cineole; eugenol; *p*-cymene; myrcene; α-terpineol; and sesquiterpene hydrocarbons.

TABLE 3
Comparison of the Physicochemical Properties of Oils of Different Basils (viz., French, Bulgarian, Indian, and American)[56,61]

Properties	French	Bulgarian	American	Indian
Specific gravity	0.8959—0.9168/15°	0.8845/20°	0.9132—0.9278°/15°	0.966/20°
Refractive index	1.477—1.488/20°	1.475/20°	1.4875/20°	1.5122/20°
Optical Rotation	−10°14′—13°52′	−13°88′	−62°1′—−9°42′	−4°54′
Acid number	0.7—3.5	0.61	0.68—1.85	2.00
Free acids%	—	0.07	—	—
Ester number	3.5—9.8	8.3	—	178.7
Esters as (linayl acetate)%	0.8—5.2	2.9	1.5	—
Acetylation number	—	190	—	259.2
Alcohols%	34.5—39.66	1.0	65.3	—
Phenols%	—	0.2	1.74	—
Solubility in 80% ethanol	1:1	1:1	1:1	—

TABLE 4
Physicochemical Properties of Kashmir Basil[53]

	Oil from	
Properties	Plant	Leaves and tops
Yield (%oil)	0.44	0.53
Specific gravity	0.877/30°	0.9640/15°
Refractive index	1.4625	1.5050
Acid value	0.2	0.7
Alcohol (linalool)%	69.4	17.7

The Bulgarian basil oil contained only traces of cinnamic acid. Geraniol, *p*-cymene and myrcene were found for the first time in basil oil. The Bulgarian basil oil does not belong to the methyl cinnamate type, it is close to the American type of sweet basil oil.

The properties of oils from different countries are given in Table 3.

7. Indian Basil Oils[53,56,62,63]

India the land of herbs and spices is rich in basil. Recently the Indian Basil has come into the field with the advancing scientific techniques of production of volatile oil.

Steam distillation of mature plants from North India yielded 0.4% oil.[56] It was a pale yellow oil with a mild pleasant odor and the physicochemical properties of which are given above under methyl cinnamate type of oil.

Baslas[53] analyzed the physicochemical and chemical properties of basil from Delhi, Kanpur, and Assam, India. The following are the main components present in the oil.

1. Delhi variety: methyl cinnamate 56.67%; *l*-linalool 4.36%; terpinene 20.85%.
2. Kanpur variety: methyl cinnamate 43.6 to 63.6%; linalool, 12 to 33.8%; methyl chavicol, 10.2 to 23.3%; ocimene 0.8 to 4.5%.

A sample of oil from the plants raised in Kashmir, India from seeds imported from France has been reported to possess the properties as shown in Table 4.

On the other hand the Assam, India variety has the following physicochemical properites:[6] specific gravity at 29.2°C, 0.9183; optical rotation −9°0′; refractive index 29.2°C,

1.4798; acid number 0.8; ester (linalyl acetate) number 7.2; solubility 1.5 to 2 vols of 80% alcohol.

The oil has been found to contain 13 components by gas chromatography. Lawrence et al.[58] have shown by gas-liquid chromatography nearly 46 components being present in the oil comprising basically monoterpene hydrocarbons, aromatic hydrocarbons, oxygenated terpenes, and sesquiterpene hydrocarbons.

The oil contains nearly 54.7% linalool and 33.9% methyl chavicol resembling in chemical composition the European type of sweet basil oil of superior quality.[2]

The oils of basil from different growing areas in India are compared in Table 2.

8. Yield of Different Components[61]

Drying in the sun and ordinary drying of basil do not affect either the yield of essential oil or the content of phenols. The best yield of eugenol from the oil is obtained by using 10% alkali solution and steam distillation. The yield of crude eugenol is 78% and that of the pure rectified product is 50% (based on the analytical content of eugenol in the essential oil). The yield of camphor is 25 to 26%. Treating the residual oil with sulfuric acid yields 18.3 to 19.9% of technical camphor.

9. Quality of Oil[6]

The oil obtained from flowers is better than the oil of whole herb in quality. The quality of volatile oil varies greatly in composition and properties owing to differences in species, soil, climatic conditions, and the part of the plant used. The European oil has the following properties reported by several authors: specific gravity at 15°C: 0.895 to 0.930; refractive index 1.477 to 1.495; optical rotation −22° to −86°; ester number 3 to 15; acid number 0 to 4 and soluble in 1 to 2 parts of 80% alcohol; alcohol content, calculated as linalool 34.50 to 39.66%. The methyl chavicol content of all these oils was about 55%, calculated by determining the methoxy number according to Zeisel's method.

10. Adulteration of Sweet Basil Oil[6]

This oil is frequently adulterated with the much lower priced Reunion basil oil. Such an addition is indicated by increased specific gravity and refractive index and by a lowered levorotation or slight dextrorotation, depending upon the amount of Reunion oil present. These discrepancies can be partly corrected by the addition of *l*-linalool; therefore, adulteration of sweet basil oil is not always detected by mere routine analysis.

11. Specifications[57]

The Essential Oil Association of America has established the specifications for volatile oil as given in Table 5.

K. USES[1,3-6,12,63,66,67,69-75]

Sweet basil is used in soups, meat pies, fish, certain cheeses, tomato cocktail, eggplant, zucchini, cooked cucumber dishes, cooked peas, squash, and string beans. Chopped basil is sprinkled over lamb chops before cooking. Basil is often used with, or as a substitute for, oregano in pizza topping, spaghetti sauce, or macaroni and cheese casserole. Basil is an important seasoning in tomato paste products in Italy.

Although not used in large quantities, the oil of sweet basil is employed quite extensively in all kinds of flavors, including those for confectionary, baked goods, and condimentary products (chili sauces, catsups, tomato pastes, pickles, fancy vinegars) and in spiced meats, sausages, etc. The oil serves also for imparting distinction to flavors in certain dental and oral products.

Sweet basil is used in seasonings of canned spaghetti sauces, meat balls, and salad.

TABLE 5
E.O.A. Standards for Oil of Basil (*Ocimum basilicum*)[57]

Property	Specifications
Preparation	By steam-distillation of the flowering tops or the whole plants
Physical and chemical constants	
Appearance and odor	Light yellow liquid with a spicy odor
Specific gravity at 25/25°C	0.952—0.973
Optical rotation at 25°C	0° to +2°
Refractive index at 20°C	1.5120 to 1.5190
Acid value	Not more than 1.00
Saponification value	4 to 10
Ester value after acetylation	25 to 45
Descriptive characteristics	
Solubility	
Benzyl benzoate	Soluble in all proportions
Fixed oils	Soluble in all proportions in most fixed oils
Mineral oil	Soluble with turbidity
Propylene glycol	Soluble up to 5% with slight haziness
Glycerine	Insoluble
Stability	
Alkali	Unstable
Acids	Unstable in the presence of strong mineral acids
Containers	Glass or aluminum containers
Storage	Store in tight full containers in a cool place, protected from light

Sweet basil or its extract, is used in combination with other spices, in flavoring confections, baked goods, puddings, condiments, vinegars, ice creams, nonalcoholic beverages, liquors, and perfumes.[5]

Sweet basil oil has its place in certain perfume compounds. Reunion basil oil is employed in all cases where the high price of the true sweet basil oil makes the use of the latter prohibitive.

The plant is considered stomachic, anthelmintic, alexipharmic, antipyretic, diaphoretic, expectorant, carminative, stimulant, and pectoral. The juice of the leaves is considered useful in the treatment of croup and is a common remedy for coughs. It has a slightly narcotic effect and allays irritation in the throat. It is used as a nasal douche and as a nostrum for earache and also for ringworm. The plant is used in homoeopathic medicine.

Oil of basil (European type) is used extensively as a flavoring for confectionery, baked goods, sauces, catsups, tomato paste, pickles, fancy vinegars, spiced meats, sausages, and beverages. It is also used for scenting dental and oral preparations and in certain perfume compounds, notably jasmine blends, to impart strength and smoothness.[3]

Reunion oil also finds similar uses. It is preferred to the European oil for use in soap perfumes. Methyl cinnamate and Eugenol types have not attained as much commercial importance.

Basil oil possesses insecticidal and insect repellent properties; it is effective against houseflies and mosquitoes. Basil was supposed to be an effective antidote to certain lizards' poison. It is also bactericidal.

Basil is diaphoretic, carminative, and stimulant; seeds are mucilagenous, demulcent, aphrodesiac, and diuretic; juice of the plant is anthelmintic. The root is febrifuge and antidote to snake poison. Juice of the leaves is dropped into the ear in earache and dullness of hearing. Mixed with a little ginger and black pepper, the leaf juice is given during the cold stages of acne. Leaves are dried and powdered and used like snuff to dislodge maggots from the nose. A 12% decoction of plant acts as a parasiticide and antiseptic and is used in nasal myosis which produces anesthesia. Basil leaves are also used for the cure of asthma.

A religious use of the leaf is found in the temples of India where the fresh leaves are floated in the holy water which is given to the devotees in small quantities for drinking.

Chapter 3

BAY

I. *LAURUS NOBILIS* (L.) — FAMILY: LAURACEAE

The bay or laurel leaf, also known as Bay leaf, is obtained from *Laurus nobilis* L., an evergreen tree belonging to the Lauraceae and Laurel family, and indigenous to countries bordering on the Mediterranean. The dried leaves of the laurel plant are called bay leaves and are not to be confused with the West Indian bay or California bay or with the Chinese variety whose leaves have a marked phenolic odor similar to the west Indian bay.[3,55,63,72,77]

A. BOTANY AND CULTIVATION[3,5,63]

The tree grows well in the subtropics and is cultivated today as a spice, or as an ornamental in the far East, the Mediterranean area, the Canary islands, Greece, Guatemala, Turkey, Russia, Mexico, France, Yugoslavia, Belgium, Central America, and milder climates of North America. It is sometimes grown in Indian gardens but does not seem to thrive well. The tree grows to a height of 50 to 60 ft. Trees should be planted either in late September or early April. They require sheltered, sunny position and will not usually survive in the northern parts of Britain and America. Bay trees are often grown in tubs which should be filled with equal quantities of sand, loam, and peat, top dressed with well-rotted manure, and never allowed to dry. The smooth bark is olive green or reddish. The leaves are smooth, lustrous, deep green on the upper surface, pale on the lower, elliptical in shape, 1 to 3 in. long tapering to a point at the base and tip of the leaf. Stem and petioles are unicellular having curved hairs. Epiderms of leaf have sinuous, thick-walled cells; stomata is in the lower epiderm only, often in pairs; mesophyl with oleoresin cavities; fiber depicted through groups of veins extending through the mesophyl. The flowers are yellow and the fruit or berry is purple-colored resembling small olives. Only the leaves are used as a spice.

Bay leaves are harvested by manual picking early in the day. To retain the natural green color of the leaves, they are dried in thin layers in trays kept in a moderately warm, sheltered place. When drying, the leaves are pressed lightly under boards to avoid a tendency to curl. Bay leaves generally are not dried in the sun, as this would cause the leaves to turn brown and lose much of the essential oil. The leaves are packed in sealed containers after about 12 to 15 days of drying.

The dried leaf remains shiny on top and rather dull underneath, where the tiny veins stand out clearly. The edge of the stiff, brittle leaf is often slightly wavy even though it has been pressed for about 10 days after having been dried in the sun. By the time it is ready for market, each leaf would have turned yellowish green in color.

The leaf has a pleasant odor, and its characteristically strong, pungent, and almost bitter flavor becomes more apparent as the brittle leaf is crushed. The taste is aromatic and bitter which intensifies within a few minutes.

B. CHEMICAL COMPOSITION[75]

Analyses of the leaf made at several locations are shown in Table 6.

C. VOLATILE OIL[76-93]

The oil of bay or laurel leaf (*L. nobilis* L.) should not be confused with that from the fruit of the same species or with bay or myrica oil from the leaves of *Pimenta acris* Kost, used for making bay-rum or again with California bay oil from the leaves of *Umbellularia californica* Nutt.[81] Lawrence[83] has reported an exhaustive coverage on the constituents of bay oil from different sources. The yield of oil can go up to 3% with the principal constituent

TABLE 6
Composition of Bay Leaves[3,63,75]

Components	Percentage
Water	4.50—10.01
Food energy	410 kcal/100 g
Protein	7.60—10.56
Fat	4.5—8.8
Total carbohydrates	65
Pentosans (%carbohydrates)	13.84
Fiber (%carbohydrates)	25.20—27.98
Ash	3.70—4.17
Calcium	1.0
Phosphorus	0.11
Sodium	0.02
Potassium	0.53
Zinc	0.04
Iron	0.53
Vitamins (mg/100 g)	
Thiamine	0.10
Riboflavin	0.42
Niacin	2.00
Ascorbic acid	46.6
Vitamin A (as β-carotene)	6185 I.U.

cineole up to 50%. The oil is a colorless liquid with a strong aromatic, camphorceas odor and a cooling taste.[5] It has a warm, fresh, penetrating, camphorlike odor resembling eucalyptus. It has a distinctly sweet, soothing, medicinal, spicy, peppery flavor with good persistence and somewhat bitter aftertaste. According to Parry,[72] laurel leaves' volatile oil have the following properties: specific gravity at 15°C: 0.915 to 0.930 (occasionally higher); optical rotation at 20°C $-15°$ to $-22°$; refractive index at 20°C 1.4670 to 1.4775; solubility 1 part in 3 parts of 80% alcohol.

Table 7 shows the characteristics of oils distilled in various places.

1. Constituents of the Volatile Oil[13,71,75]

The volatile oil consists of esters $C_{10}H_{17}.OH$ acetate (about 13%), total alcohols $C_{10}H_{18}O$ (about 23.2%), free alcohols (about 13%), cineol (45%), terpeniol and geraniol (18%), acetic ester (13%), free eugenol (0.53%), eugenyl acetate (1.1%), methyl eugenol (3%), terpene hydrocarbons (12%), and principally pinene, phellandrene, and sesquiterpenes (3 to 4%).

The monoterpene hydrocarbons consist of α-pinene (2.8%), β-pinene (traces), α-phellandrene (2.9%), myrcene (70.0%), β-phellandrene (70.2%), β-cymene (1.0%), ocimene, α-thujone, 3-carene, sabinene, α-terpinene, γ-terpinene, and terpinolene, and as has been reported by Ikeda et al.[79] by silicic acid chromato-strips and gas-liquid chromatography and by infrared spectra. These workers obtained α-terpinene, γ-terpene, and terpinolene by silica gel and isomerization of limonene.[79]

According to Movesti[77] the Italian oil contains *d*-α-pinene, phellandrene, cineole (35 to 45%), terpineol, *d*-linalool, geraniol, eugenol, acetic, isovaleric and caprioc acids and isobutyric acids.[77] Phenyl urethan has also been identified in the oil. Pentosans are present up to the extent of 13.84% on a dry matter basis.[75]

The high boiling fractions of the oil on subjecting to gas-liquid chromatography are shown to contain *d*-linalool (7.90%), 1—4 formate (17.90%), methyl eugenol (7.20%), acetyl eugenol (0.10%), and sesquiterpene hydrocarbons (3.70%) which include β-caryophyllene, β-sabinene, γ-cadinene, and at least five additional unidentified carbonyl containing compounds.

TABLE 7
Physicochemical Properties of the Oil of *Laurus nobilis* L. from Different Sources[74,81]

Source (Values in parenthesis are yield of volatile oil)	$d_{15°}$	α_D	$n_{D,20}$	Ester number	Ester number after acetylation	Acid number	Solubility in 80% alcohol
German and French	0.915—0.932	−15°—18°	1.469—1.477	28—43	58.3—77.7	Up to 2.0	1.3 vols
Flume	0.9211	−13°52′	1.47156	31.9	—	1.6	2.5 vols of 70% alcohol
Corfu	0.9177—0.9211	−21°40′	1.47107—1.46862	29.8—43.8	36.2—82.5	Up to 1.5	1 vol
Cyprus	0.9405	−5°40′	1.47372	25.3	95.6	1.6	2—3 vol of 70% alcohol
Palestine	0.916—0.924	−14°— −20°	1.46516—1.46575	21—49	43.2—81.4	Up to 2.2	1—2 vols
Asia Minor	0.9268	−15°50′	1.46565	34.8	60.1	0.8	1 vol
Dalmatia	0.9268	−14°36′	—	—	—	—	2.5 vols of 70% alcohol
Syria	0.9161	−14°20′	—	—	—	—	1 vol
Italy (1.28%)	0.921	−18°	1.471	2	37	84	1 vol
Crimea (0.67%)	0.917	−16°	1.471	1.7	31	52	—
Ssuchum (1.1%)	0.916	−18°	1.464	0.7	37	51	—
Ssotschi (0.56%)	0.916	−18°	1.468	0.4	43	63	—
Italian (2%)	0.920—0.930	−13°—21°	1.465—1.480	—	—	—	1— 1.5 in 3 vols

The terpeneless oil can be prepared by using silicic acid as absorbent and petroleum ether as solvent.[72]

D. USES[1-6,12,16,55,70,74,76]

Laurel leaves enjoy wide culinary use. They are employed to flavor, meat, game, poultry, and fish dishes, soups, and sauces. Cracked laurel leaves are an ingredient of whole mixed pickling spice. Oil of bay and bay leaves are used in the preparation of pickles and as spice in flavoring of vinegar.[3] Bay leaves, bitter, spicy and pungent in taste, are popular French cuisine, used for flavoring bouillabaisse, bailles, meats, fish, poultry, vegetables, and stews.

Leaves and berries of bay tree are reputed to be astringent, stomachic, aromatic, stimulant, and narcotic. Decoction of leaves is used in leucorrhea, diseases of the urinary organs, and dropsy. The leaf decoction is a powerful emmenagogue. The seed oil is used to remove rheumatic pains.[3]

Bay leaf lends a unique aroma to calves liver, lamb, meat loaf, fish, poultry, soup stocks, and all stews. A few grains of the ground spice added to tomato soup or spaghetti sauce perk up the flavor and aroma of the dish. The ground spice may also be added to marinades, pot roasts, stews, ragouts, game dishes, fish sauces, homemade pickles, cream sauces, onions, and squash.[5]

Both leaves and fruits possess aromatic, stimulant, and narcotic properties and were formerly employed for hysteria, amenorrhea, and flatulent colic. Externally, however, commercial oil of laurel berry is sometimes applied as a stimulant in sprains, but its principal use is in veterinary medicine.[6] Berries are also used in diarrhea and dropsy.

Chapter 4

BERGAMOT

I. *MONARDA FISTULOSA* (LINN) (WILD BERGAMOT) — FAMILY: LABIATAE

Bergamot or wild bergamot is the species native to the Western Hemisphere and belongs to the mint, Labiatae family. It is commonly called wild bergamot. It is mostly found in the western states north to British Columbia and across the continent to Quebec and also some parts in North Carolina (U.S.).[3,6,12,13] Red bergamot is also known as fragrant balm, bee balm, red balm, Indian plume, or Oswego pea, after the Oswego Indians who were using it abundantly. Lemon bergamot also originates from North America and has aromatic properties similar to the wild and red bergamots.

A. BOTANY[3,51,72]

The plant is a beautiful perennial with soft, cotton-like foliage and dense clusters of purple flowers with brilliant red corollas. It grows 2 to 3 ft. tall. Its leaves have pleasantly scented lemon odor. It flowers during the month of August.

B. VOLATILE OIL[3,13,72,74,94]

Distilling the flowering fresh herb yields 0.3 to 1.0% of oil whereas the dried flowers and leaves yield 2.7 to 3.1%. The later flowering period flowers yield 0.5 to 0.6% oil. The odor of these oils is more phenolic and carvacrol-like.

1. Constituents[3,52,74]

The chemistry of "wild bergamot" oil (*Monarda fistulosa*) has been investigated and the following compounds reported: limonene, *p*-cymene, carvacrol (67%), thymohydroquinone, thymoquinone, dihydroxy thymoquinone, *d*- and *l*-α-pinene, two other terpenes (not further identified), butyric aldehyde, isovaleric aldehyde, piperonal (?), dihydrocuminyl alcohol, acetic, butyric, valeric, and caproic acids, geraniol and linalool (?), and quinones. The physicochemical properties of volatile oil are shown in Table 8. The water distillate of the leaves contains, acetone, formaldehyde, methyl alcohol, formic acid, and acetic acid.

Carvacrol is the main constituent (about 67%). It is identified as dicarvacrol m.p. 147 to 148°c; carvacrol sulfonic acid m.p. 58 to 59°C; nitrosocarvacrol m.p. 153 to 154°C; and dinitrocarvacrol m.p. 117 to 119°C.

The chemical composition of bergamot oils from Corsica, Italy, and the Ivory Coast are compared.[95] Summary of the results with percent ranges in parentheses are α-pinene (0.76 to 1.56); camphene (0.02 to 0.04); β-pinene (3.5 to 7.5); sabinene (0.69 to 1.26); myrcene (0.69 to 0.90); α-phellandrene (0.04 to 0.09); limonene (26.24 to 35.64); γ-terpinene (4.14 to 8.97); ρ-cymene (0.08 to 0.34); terpinolene (0.17 to 0.21); nonanal (0.03 to 0.07); citronellal (0.02 to 0.06); decanal (0.06 to 0.09); linalool (12.67 to 26.37); linalyl acetate (31.11 to 32.71); terpinen-4-ol (0.21 to 0.29); bergamotene (0.29 to 0.32); neral (0.11 to 0.30); α-terpineol (0.08 to 0.16); geraniol (0.25 to 0.58); bidabolene (0.56 to 0.77); geranyl acetate (0.16 to 0.26); nerol (0.05 to 0.33); and geraniol (0.02 to 0.05).

Lamparsky[96] examined the chemical composition of bergamot oil. The major constituents identified with percentages in parentheses were α-pinene (1.0); β-pinene (5.7); sabinene (1.0); myrcene (0.9), limonene (33.0); γ-terpinene (3.7); *p*-cymene (2.7); α-bergamotene (0.23); caryophyllene (0.27); β-bi-abolene (0.57); linalool (13.45); linalyl acetate (31.30); nerol (0.10); neryl acetate (0.42); geraniol (0.05); geranyl acetate (0.46); α-terpineol (0.13); and α-terpinyl acetate (0.21).

TABLE 8
Physicochemical Properties of Volatile Oil from *Monarda fistulosa* as Studied by Different Authors and from Different Places[56,74,86]

Properties	Cited in 56	Cited in 56	Texas	Colorado	Central U.S.
Specific gravity	0.916—0.941	0.9382/20°	0.929/15°	9.939/15°	0.924/15°
Optical rotation	Slightly levorotatory	—	Too dark	Too dark	Inactive
Refractive index	—	1.5000	1.5018	—	—
Phenol content (as carvacrol)	52.0—72.0% (mostly carvacrol)	55.4%	40.0%	64.5%	65.0%
Solubility	—	—	Turbid in 10 vol of 70% alcohol	Opalascent in 2.5 vol of 70% alcohol	Soluble in 2.5 vol and more of 70% alcohol

TABLE 9
Physicochemical Properties of *Monarda punctata* Oil as Reported from Different Workers and Places[13]

Constants	Cited in 56	Florida	Texas
Specific gravity at 15°	0.930—0.940	0.949	0.941
Optical rotation	Inactive	−0°16′	+0°25′
Refractive index at 20°	—	1.5088	1.5055
Total phenol constant	60—70%	—	61.0%
Thymol content	44—61%	71.5%	—
Solubility at 20°	—	3 vol of 90% alcohol	2.5 vol and more of 70% alcohol

In addition, the same authors identified a number of trace components which were found in the oil for the first time. These compounds are *cis*-1,2-epoxylimonene, *trans*-1,2-epoxylimonene, 2,3-epoxy-*cis*-ocimene, 2,3-epoxy-*trans*-ocimene, humulene epoxide I, humulene epoxide II, isocaryophyllene oxide, terpinen-4-ol, *cis*-sabinene hydrate, *trans*-sabinene hydrate, *cis-p*-menth-2-en-l-ol, *trans-p*-menth-2-en-l-ol *p*-mentha-1,8-dien-9-yl-acetate, *p*-mentha-1,3-dien-7-yl-acetate, p-menth-1-en-9-ol, *p*-mentha-1,8-dien-9-ol-hotrienol, 2,3-epoxy-neryl acetate, 2,3-epoxy-geranyl acetate, terpinen-4-yl-acetate, hotrienyl acetate, *p*-menth-1-en-9-yl-acetate, nerolidol, α-bisabolol, β-bisabolol, α-cadinol, caryophyllene alcohol, caryophyllenol I, caryophyllenol II, and β-photosantalol A.

II. *MONARDA PUNCTATA* (LINN) (HORSEMINT)[3,13,16,52,70,72]

Monarda punctata L., the so-called horsemint, grows wild on light sandy soil in many parts of the U.S., especially in the Southeast, also in the Midwest, from Wisconsin and Kansas to Texas and Florida, and in some parts of Europe.

A. VOLATILE OIL[3,13]

Distillation of fresh plants yields up to 1.0% of oil and dried plants yields about 3.0% of oil. The physicochemical properties of volatile oil from *Monarda punctata* are given in Table 9.

The odor and flavor of this oil were quite similar to those of oil of thyme.

B. CHEMICAL COMPOSITION[3]

Thymol the main constituent of oil of *Monarda punctata* was identified over a century

TABLE 10
Range of Physicochemical Properties of Volatile Oil from *Monarda menthaefolia*[13,56]

Constants	Values
Specific gravity	0.932, 0.9666—0.952
Specific rotation	+1°33′—+1°34′
Refractive index at 25°C	1.5110—1.5143
Phenol content	82—93.2%

ago. Many years later the presence of the following constituents has been established: *d*-limonene, *p*-cymene, thymol (about 61%), thymohydroquinone, linalool formaldehyde, and acetaldehyde.[3]

The Texas Monarda oils (*M. punctata* spp. Stanfield) contains phenols in the oil which mainly consist of carvacrol, and those in the oils of *M. punctata*, spp. typica and immaculate consist predominantly of thymol.

C. USES[1-6,12,16,70,74,56]

Oil of *Monarda punctata* L. has not yet attained any wide commercial importance, but as has been pointed out, the oil could be used in the place of thyme oil, provided it can compete with the price of the latter.

III. *MONARDA CITRIODORA* (CERV.) (LEMON MINT)[3]

Monarda citriodora cerv; the so-called lemon mint, grows wild in the U.S., particularly in the section roughly bordered by Illinois, Nebraska, and Texas.[56]

A. VOLATILE OIL

Distilled flowering dried herb, yields about 1% of oil and the following properties were reported:[56] specific gravity at 20°: 0.9437 to 0.9603; phenol content 65.0 to 80.0%; and aldehyde content 1.2 to 4.0%. The oil has a carvacrol and lemon-like odor.

B. CHEMICAL COMPOSITION[13,66,74]

The following constituents are reported: carvacrol, thymol, thymohydroquinone, citral, and *p*-cymene(?). The oil of *M. citriodora* cerv. distilled from plant material collected near Texas consisted of predominantly thymol.

C. USE

The oil has not attained much commercial importance.

IV. *MONARDA MENTHAEFOLIA*[3,13]

The herb *Monarda menthaefolia* grows wild in Colorado and Wyoming.

A. VOLATILE OIL[1,13]

The flower heads yield 0.31%, the leaves 0.32%; the stems 0.05% and the roots only traces of oil. The oil has a similar odor of *Monarda punctata* L. The physicochemical properties of volatile oil in *Monarda menthaefolia* are given in Table 10.

B. CHEMICAL COMPOSITION

The oil contains the following constituents: *p*-cymene, carvacrol, hydrothymoquinone, geraniol, linalool, and acetic acid.

C. USE

The oil has not attained much commercial importance.

V. *MONARDA PECTINATA* (NUTT)[13]

The fresh flowering herb of *M. pectinata* Nutt, when distilled yields 0.5% of volatile oil which has these properties: specific gravity at 25°: 0.9496; optical rotation too dark; refractive index at 25°: 1.5070, and phenol content 77%. The phenols contained in the oil consist chiefly of carvacrol which is identified by the preparation of its nitro compound with an m.p. 153°C.

VI. USES OF BERGAMOT LEAF[1-6,12,16,52,74,97]

The dried leaves make a delightful drink and bergamot tea may be served in place of other teas. The fresh leaves and tender sprigs make a delicious addition to tall wine drinks and lemonades. The leaves may also be put into soft fruit drinks and iced cups. The fresh tea from bergamot leaves are known to have relaxing and soporific value.

Chapter 5

BORAGE

I. *BORAGO OFFICINALIS* — FAMILY: BORAGINEAE

Borage, native to Asia Minor, Greece, Italy, the Mediterranean coasts, Persia, and Sicily, grows in many parts of the world. In North America, it is found chiefly in the eastern states. The tender young leaves of the borage plant taste fresh and cool and have a flavor of cucumber. Borage has been cultivated in England for so many generations it is difficult to decide whether it is a native herb or not. It also grown in hill stations in India.[70,71]

A. BOTANY[3,12,70,69]

It is a sturdy annual herb and is one of the most beautiful plants in spite of its coarse foliage with its stringing, hairy surface. The plant grows to a height of 2 ft or more and borage flowers form drooping clusters of heavenly sky blue five-pointed stars. Borage can be grown with ease from seed in all temperate or warm climates and prefers a well-drained calcareous soil and sun.

The leaves are large ovals of grayish green color covered with stiff hairs. They contain a volatile oil having aromatic odor and the extractive of the leaves tastes like cucumber.

B. CHEMICAL COMPOSITION[3,51,56,74,98,99]

Borago officinalis contains protein (12.38 to 14.14%), fat (3.14 to 5.74%), and ash (15.94 to 21.92%); and small quantities of allantoic acid. The changes in leaf during the growth of the borage plant reveal that as the maturity progresses the ash content is reduced from 16.13-48.23% to 8.82-12.73%. It rises to 16.31 to 23.06% during the period of actual fruit formation; the cellulose content is comparatively low but increases towards the fruit bearing period. Composites are characterized by a comparatively high fat content (3.14 to 5.74%) and protein especially during the period of fruit bearing. Compared with annuals of other families, composites contain more nitrogen-free extract substances but ash content is lower.[56]

C. USES[1-6,16,67,74,56,99]

The dried leaves of borage may be used as well as the fresh to flavor soups and stews in place of parsley. A few fresh leaves tossed with a green salad are delightful and when cooked with other vegetables they add a soothing taste. This is especially so with green peas, beans, and oyster plant. Borage is placed in water in which the vegetables are cooked.

In Western Europe young borage leaves are cooked as Americans or Indians do with spinach, and the English often mix the herb with other greens and serve it plain or creamed as a principal vegetable. For use in salads, the leaves must be chopped finely as their hairy texture makes then unappetizing. Borage with yogurt is delicious. In parts of Italy where borage grows in wild profusion the leaves are used as a stuffing for ravioli or boiled as a spinach or even fried in batter.

The dried leaves are particularly aromatic and can be used alone or blended with other herbs to prepare tea. Flowering tips and leaves give delightful flavor to many iced drinks. The flowers may also be candied or crystallized and then eaten as a confection or used to decorate cakes and cookies. One or two borage floatings in a cup of wine improve the flavor of the drink.[3]

In its medicinal use, a poultice of borage leaves is known as an excellent application for swellings or for inflammation of any type. The emollient virtues which are possessed by all parts of the herb make it valuable in coughs and other complaints of the air passages and it has been recommended to tubercular patients also. In each case the infusion should be prepared and taken in doses of a wineglass full every 3 hr or whenever the cough is troublesome. Borage has recently been found to be cyanogenetic.

Borage is generally available fresh in markets. It wilts immediately after it is picked. Its use as a spice is increasing quickly.

Chapter 6

BURNET

I. *SANGUISORBA MINOR* — FAMILY: ROSACEAE

The salad burnet or garden burnet was an essential garden herb in Elizabethan England and was taken to America by the early settlers.

A. BOTANY[3,51,70,74]

Burnet or salad burnet is a hardy perennial with several species. Two belong to the Rosaceae family, and one to the carrot, Dancus family. The species of salad burnet known as the *Sanguisorba minor* is widely cultivated in the U.S. especially in North Carolina. In Southern Europe and Western Asia, The Great Burnet, *Sanguisorba officinalis*, grows wild.

All burnets look alike; and the attractive, lacy, light green foliage makes a lovely border plant in any herb garden. The plant forms small mats or clumps of pinnate, feathery, foliage about 8 in. long and the rose colored or white flowers grow in flat umbel at the end of the tall stems.[3] Fresh leaves are practically available all year from herb gardens. It is a small perennial with rather characteristic little toothed leaves growing in pairs. The stem grows to nearly 1 ft high. The flower head is comprised of greenish bobble feathered with long purple red stamens. It is also easy to cultivate in the garden. It grows in practically any soil. It is generally raised from seed or by division of clumps.

The tender young leaves have a distinct, delicate flavor, resembling that of cucumbers.

B. CHEMICAL COMPOSITION[3,56]

The polyphenol components of the extracted tanins include hydroquinones (most active), resorcinol, catechol, pyrograllol, and gallic acid, all of which are both bacteriostatic and bactericidal to the dysentery group.

C. PACKING

Dried leaves are available in 1/2 to 1 ounce containers and in bulk at drug stores and also at herb dealers.

D. USES[1-6,52,70,74]

Both the dried and the fresh leaves may be used in preparing herb teas and herb vinegars. The delicate flavor of the young tips of the salad burnet is especially delicious in all vegetable salads and the dried leaves are just as useful as the fresh, for they may be blended in with the salad dressings. A spring of the delicate looking foliage is a most attractive addition to the appearance of an iced drink as well as its flavor. It is also a classical ingredient in several butters and sauces, such as ravigote and chivy, where it finds use as finely chopped fresh leaves. Usually borage is used for beet, cabbage, carrot, celery, lettuce, mixed green or tomato salads, and asparagus, lima bean, or mushroom soups.

Chapter 7

CAMOMILE

I. *ANTHEMIS NOBILIS* — FAMILY: COMPOSITAE

Camomile is both sweet and wild; German camomile grows profusely in Europe and temperate Asia as well as in the British Isles and the U.S. The sweet variety is a perennial, but the wild one is an annual. It is easy to grow both of them in the garden and for commercial purposes. Not much is known to definitely say which of these plants is true camomile or which of the two is more effective.[3,62]

A. BOTANY[3,62]

Camomile (or chamomile) is an annual or perennial herb. Leaves are alternate, deeply toothed, or pinnatisect. Heads are terminal peduncle rarely corymbose, heterogamous, and radiate (very rarely disc-form); flowers are female, fertile, sterile, or neuter; tube aterete or two winged, leaf spreading; disc-flowers hermaphrodite, fertile tube compressed or sometimes two winged; base usually 1 to 2 gibbous, involucre hemisphere; bracts many seriate, appressed, rigid, margins searious outer shorter, receptacle convex or elongate; pales narrow rigid or hyaline, sometimes embracing the flowers. Anther bases are obtuse. Style arms are hermaphrodite with truncate penicillate tips. Achnes are oblong, glabrous, 4 to 5 angled, ribbed or many striate, truncate; pappous very short, and paleous are of membranous and large.

The plant is a hardy European perennial which is grown chiefly in herb gardens. The German camomile or *Matricaria chamomilla* is an annual which resembles the Roman camomile and the fragrance of the flowers is warm and sweet. Flowers are used mostly for fragrance rather than leaves. It is a low plant which spreads and creeps as it runs along the ground. Its white daisy-shaped flowers grow upward on long thin stems and the herb blossoms all summer long until the frost arrives. The dark green pinnate foliage is about 2-in. long and smells is like that of sweet ripe apple.

B. SPECIFICATIONS[57]

The specifications and standards for the oil of English Chamomile are given in Table 11.

C. VOLATILE OIL

Nano and co-workers[100,101] showed that oil of Roman camomile (*Anthemis nobilis* L.) contains α-pinene, β-pinene, β-copaene, δ-cadinene, pinocamphone, *trans*-pinocarvacol, myrtenal, hexylacetate, 2-methylbutyl butyrate, 2-methylpropyl-2-methylbutyrate, 2-methylbutyl 2-methylbutyrate, 2-methylpropyl angelate, butyl angelate, 3-methylbutylangelate, and 3-methylpentylangelate. The same authors compared the composition of three varieties of Roman camomile.[101] These authors also found that the oils contained α-pinene (<0.5 to 10.0%), camphene (0 to <0.5%), β-pinene (0 to 10.0%), sabinene (0 to 10.0%), myrcene (0 to <0.5%), 1,8-cineole (<0.525%), γ-terpinene (0 to <0.5%), *p*-cymene (0 to <0.5%), α-*p*-dimethylstyrene (0 to <0.5%), copaene (0 to <0.5%), β-copaene (0 to <0.5%), caryophyllene (0 to 10.0%), δ-cadinene (0 to 0.5%), hexylacetate (0.5 to 10.0%), myrtenal (<0.5 to 10.0%), 2-methylbutyl 2-methylpropionate (0.5 to 25%), 2-methylpropyl butyrate (0.5 to 10.0%), 2-methylbutyl 2-methylbutyrate (0.5 to 25%), 2-methylpropyl 3-methylbutyrate (0.5 to 10.0%), propyl angelate (0.5 to 10.0%), 2-methylpropyl angelate (0.5 to >2.5%), butylangelate (0.5 to 10%), 3-methylbutyl angelate (10 to 25%), and 3-methylpentyl angelate (<0.5 to 10%).

TABLE 11
E.O.A. Standards for the English Camomile Oil *(Anthemis nobilis)*[57]

Properties	Specifications
Preparation	By steam distillation of the dried flowers
Physical and chemical constants	
Appearance and odor	A light blue or light greenish blue liquid with a strong aromatic odor, characteristic of the flowers, with age the color may change to greenish yellow or yellow-brown.
Specific gravity at 25°C	0.892—0.910
Refractive index at 20°C	1.4400—1.4500
Acid value	Not more than 15
Ester value	250—310
Solubility in alcohol	Soluble in 2 and more vols of 80% alcohol
Descriptive characteristics	
Solubility	
Benzyl benzoate	Soluble in all proportions
Diethyl phthalate	Soluble in all proportions
Fixed oils	Soluble in all proportions in most of the fixed oils
Mineral oils	Almost completely soluble
Propylene glycol	Soluble with a slight haziness
Stability	
Acids	Fairly stable to weak acids
Alkali	Unstable in the presence of alkali due to saponification
Containers	Shipped preferably in glass or aluminum containers
Storage	Stored in tight, full containers in a cool place, protected from light

II. *ANTHEMIS GAYANA* (BOISS)[3,62]

This is a small tomentose canescent herb with stems filiform, erect, very short, scapiform, petiolate, in outline ovate or oblong, pinately or subpinately divided. Peduncles are elongate and finally slightly thickened. Heads are small. Outer leaves are involucre lanceolate, somewhat acute, inner ones oblong, searious at the apex. Receptacle is ovate, pales lanceolate carinate, acuminate, shorter than the disk. Ligules are white ovate slightly shorter than the disk, and hairy at the base. Achenes smooth, sulcate, slightly narrowed at the base, angular, truncate at the apex. The margin is acute and narrow. Mostly it is grown in Baluchistan and Persia.

A. USE[1-6,70,71,74]

Today all over Europe, camomile tea is widely used in place of regular tea and coffee. Even in North America, the dried flower heads are used chiefly in herb teas. The oil is used for cordials, perfumes, flavoring, infusions and medicines. The leaves are known to be eaten to cure pains in the chest.

Chapter 8

CHERVIL

I. *ANTHRISCUS CEREFOLIUM* L. (HOOFM) — FAMILY: UMBELLIFERAE

Chervil, is a small, low-growing annual of the parsley family. Distinguished by the bright, light-green color of its lacy and fernlike foliage, it is quite similar in appearance to parsley, although more delicate.[3]

Chervil, is a native of southern Russia and western Asia. It is also cultivated in France and England, and recently has also been grown commercially in the U.S., principally in California. It grows well in temperate zones. The chervil plant grows 8 to 16 in. high and resembles parsley in its habit of growth. It has a delicate curly, feathery leaves, with small white flowers. Leaves may be harvested about 90 days after sowing, and leaves should be picked first (as in the case of parsley). It must be harvested frequently enough to avoid flowering, thus encouraging a new leaf crop. The plant is almost cut to ground level while harvesting.[3,70]

Sweeter and more aromatic than parsley, chervil has been called the ('gourmet's parsley'). It has a delightful anise-like fragrance with a slight tint of pepper flavor.[3]

To retain the desired green color, the leaves are dried at a moderate temperature of about 90°F.[52]

The chemical composition of chervil is shown in Table 12.

A. USES[1-6,52,70,74,78]

In France and Switzerland chervil is used extensively to flavor salads, sauces, and stuffings for poultry, fish, and shellfish. The herb is chopped fine and sprinkled over fish, chicken, egg dishes, soups, and salads. Dried leaves are used in stuffings. This lovely green herb is also used in soups, stews, and omelettes much as parsley is used, either fresh or dried. The fresh leaves also make a beautiful garnish for salads. A tuberous-rooted variety of chervil is grown and eaten as carrots. The leaves are sometimes used as garnish in the same way as parsley and are floated on the top of certain soups. Their delicate and diminutive leafage makes an extremely attractive and appetizing addition. In French cuisine it is usually included in the combination of three or more herbs ground fine and carefully blended and known as ''fines herbes'' (other herbs that may be used in fines herbes are parsley, tarragon and chives, sage, savory, and basil). Chervil is mixed with tarragon and chives to make *ravigote*, a savory sauce. Its tendency to bring out the flavor of other herbs makes chervil especially prized as an aromatic seasoning and supplement. The leaves are reported to be diuretic and stomachic.

TABLE 12
Chemical Composition of Chervil[5] (100 g Edible Portion)

Components	Quantity
Water	7.2%
Food energy	273 kcal
Protein	23.2%
Fat	3.9%
Total carbohydrate	49.1%
Fiber	11.3%
Ash	16.6%
Calcium	1346 mg
Iron	32 mg
Magnesium	130 mg
Phosphorus	450 mg
Potassium	4.74 g
Sodium	83 mg
Zinc	9 mg

Chapter 9

CHIVES

I. *ALLIUM SCHOENOPRASUM* — FAMILY: LILIACEAE OR AMARYLLIDACEAE

The chive a ''little brother'' of the onion is a small grass-like hardy perennial preferring the cooler climate of the Northern Hemisphere. Chives require an abundance of moisture and a light soil with adequate humus. They are grown more in Scandinavia, Germany, and England than in Southern Europe of the Mediterranean region.[3] In India it is grown as a garden crop. They are also grown in Canada and the northern U.S. Chive[5] is raised in many parts of the world as a vegetable or herb of the home garden where its small size and great adaptability make it more useful than the common onion.

A. BOTANY[3,52,60]

Chive usually grows to a height of 6 to 12 in. The young tender, pencil-shaped tubular leaves of this dark-green lavender-flowered herb possess a light, delicate onion flavor; it produces numerous stalks with globose heads of purple flowers amid an abundance of upright but flat hollow leaves. Chives are perennial, easy to cultivate, and will grow in any garden soil.

B. CULTIVATION[3,52,60]

They can be grown from seed, but the more usual practice is to divide the large clumps in spring or autumn, leaving five or six little bulbs together, planting these smaller clumps about 9-in apart. They will spread very quickly and can even be divided again the following year. The leaves will get coarse if the plants are neglected. In any case, the chives should be well weeded, the old leaves cut off as they die, and care should be taken to see that all flowering shoots are removed as they reappear for if they are allowed to bloom the leaves will be impoverished. Chives are completely immune to cold, can withstand drought, and grow well in a wide variety of soils.

C. HARVESTING[3,60]

Chives are cut 3 to 4 times a year 2 in. above ground. The first cutting and curing of the leaves take place 5 or 6 weeks after spring planting and before flowering. Only the green tops or leaves of the chive are eaten, not the small, poorly developed bulbs.

D. PROCESSING[3,60]

Chives cannot be dried by normal kitchen means. Commercial production of chives for quick freezing and freeze drying is in vogue. The fresh frozen product is used by dairies for flavoring and garnishing cottage cheese. The freeze-dried chives are available in packages for garnishing and culinary purposes. During the past 10 years, freeze drying has substantially increased the demand in the U.S. for this previously little known condiment.

E. COMPOSITION OF CHIVES

The composition of chives is given in Table 13. A γ-glutamyl tripeptide isolated from seeds of *Allium schoenoprasum* is identified as an *S*-(propene-l-yl)-L-cysteine derivative.[3]

F. SPECIFICATIONS (COMMERCIAL)[1-6]

The moisture content should be between 1 and 3%. One pound of dried leaf is equivalent to nearly 12 lb of fresh chive. All pathogens must be completely absent. The standard plate

TABLE 13
Composition of *Allium schoenoprasum*[11,12,81,102-106]

Components	Quantity (%)
Water	92.2
Protein	2.6
Fat	0.33
Nitrogen-free extract	3.09
Fiber	1.48
Ash	1.28
Calcium	0.048
Iron	0.0084
Phosphorus	0.057
Sugars	3.5
Ascorbic acid (mg%)	52—97
Oxalic acid	0.1
Antithiamine active substances	0.5

count shall not exceed 20,000/g. The coliform count is not more than 20/g. The product must be bright green and free from yellow pieces and uniformly sized with approximately one million pieces to a pound.

G. USES[1-6,16,74,81,87,102]

Chives are practically always used fresh. However, in the recent development of numerous herb salts, dried or powdered leaves of chives are among those used in combination with common table salt and other herbs. The herb salt forms a seasoning for those who wish the convenience and the new taste thrill of a quick seasoning. The fresh green leaves are used to flavor all foods in which a mild onion flavor is desired. The bulbs may be pickled as tiny onions and have an infinitely more delicate flavor. Cream cheese mashed with chives, as an appetizer is a good dish. Creamed cucumber with chives will be excellent. In France, chives are added to egg preparations also. Fish pie with chive is supposed to be delicious. Dishes like chive cheese omelette, Eggs Benedict with Hollandaise chive sauce, and chicken chive fricassee are usual in some countries like France and Switzerland. Freeze-dried chives may be used in prepared soup mixes, salad dressing, cocktail dips, sour cream, and cottage cheese products.

Chive oil from *Allium schoenoprasum* has a future as an antibiotic, since it inhibits *Myocobacterium tuberculosis* including a streptomycin-resistant strain *in vitro* at 1.25 to 2.5 γ/ml.[102] Other species of microorganisms require 200 γ/ml or greater for inhibition.[102]

Chapter 10

CORIANDER

I. *CORIANDRUM SATIVIUM* L. — FAMILY: UMBELLIFERAE

The earliest reference to coriander can be traced as far back as 5000 B.C. in Sanskrit writings in India. Later the presence of coriander was discovered during 900 to 1100 B.C. in Egyptian tombs showing the ancient heritage of this spice. It is said to be spread to the Mediterranean region also. It is widely grown and used in India, the U.S.S.R., Hungary, Poland, Romania, Czechoslovakia, Guatemala, Mexico, and the U.S. India has been one of the major producers of this spice. Practically all the parts of the plant, that is, tender stem, leaves, flowers, and fruits have a pleasant aromatic odor and are widely used. It is essential to distinguish sharply between the green leaf and the ripe seed in coriander plant. The taste is absolutely different. The green coriander leaf can be recognized in places such as haymarkets not only from the smell, but because the lower leaves are typically fanlike and the upper leaves feathery.[1-6]

A. BOTANY[5,6,12]

The plant is a strong odiferous hardy annual of the parsley family. It thrives best in weedless, fertile, and deep soil. It is preferably grown under irrigation. The plant has many branches and leaves are serrated. It grows sometimes to nearly a foot tall and produces a slender, erect, hollow stem with small, light pink flowers in a compound umbell. It is not an herb to dry but may be quick-frozen and preserved with salt in oil.[5,6,12]

B. COMPOSITION OF LEAVES[18]

Coriander leaves have the composition as shown in Tables 14 and 15.

C. ESSENTIAL OIL FROM LEAF[5,6]

Stalks and leaves of coriander yield considerable volatile oil (about 0.1 to 0.6%) but the leaf oil is not used instead of seed oil in any flavor, perfume, or pharmaceutical formula. The leaf volatile oil has been shown to have the odor of decylaldehyde and other higher fatty aldehydes.

The properties of coriander herb oil from different sources are tabulated in Table 16.

D. USES[3,4,6,12,71,74]

The fresh stem, leaves, and fruits of coriander are all used in many culinary preparations and have a specific aromatic odor. The entire plant when tender is used in India in preparing chutneys and sauces and is used for flavoring curries and soups. The young leaves are also used to flavor soups and salads in Spain, Mexico, South America, the Mediterranean and in North American dishes. It is generally used either as whole leaves for dressing or chopped in soups, curries, and chutneys. The flavor of green coriander enhances the taste of some of the Indian dishes where green chillies are used. The leaves are getting more popular every day especially in western dishes.

TABLE 14
Composition of Coriander Leaves (Fresh)[18,103-105]

Components	Quantity (%)
Moisture	86—88
Protein	3.3
Fat	0.6
Carbohydrates	6.5
Mineral matter (total ash)	1.7
Phosphorus	0.06
Iron	0.02
Vitamin A	10,460 I.U./100 g
Niacin (mg%)	0.8
Vitamin B_2 (mg%)	60
Vitamin C (mg%)	150—200

TABLE 15
Composition of Green Young Coriander Plant (Moisture-Free Basis)

Content	Values (%)
Herb oil (volatile%)	0.1—0.95
Crude fiber	27.7
Calcium	1.23
Boron (ppm)	25
Strontium (μg/g)	857
Available carbohydrate	10.6
Oxalic acid	0.09
Vitamin A (μg/100 g)	5200
Vitamin C (mg/100 g)	98 to 250

TABLE 16
Properties of Coriander Herb Oil from Different Source[56,109-111]

Properties	U.S. oil (Ohio)	Indian oil	U.S.S.R. oil
Specific gravity at 25°C	0.849	0.849	0.8524
Optical rotation α_{20}^{D}	3°32′	+2°1′	+2°33′
Refractive index	1.4540	1.4548	1.4555
Saponification number	22.8	50.1	—
Aldehyde content (%)	65.5	72.0	95.0
Acid number	—	—	1.85
Solubility at 25°C in 20% alcohol	3.50%	1.50%	Insoluble

Chapter 11

COSTMARY

I. *CHRYSANTHEMUM BALSAMITA*—FAMILY: COMPOSITAE

Costmary is a tall, luxuriant perennial of the Compositae family, native to western Asia. It probably comes from Kashmir in India.[3] Costmary is also known as alecost since at one time it was used to flavor ales and beer. Another common name is bible leaf, from the custom of placing one of the leaves in the Bible as a book mark.

The plants produce large clumps of long, light green, slender leaves and the flowers look like yellow buttons or daisies with yellow centers and a few straggling white petals. Costmary grows easily in most soils. The herb has a very agreeable, minty scent, while the flavor of the leaves is somewhat bitter. After being dried and infused, the flavor changes to a slightly lemon-like flavor. Dried leaves in bulk are available from herb dealers and herb gardens.[70,72]

A. USE[3,70,72]

The fresh, tender leaf is used to flavor certain meats and poultry. The flavor is very dominant, so care should be taken not to use more than a leaf, especially for cakes and jellies. It is also used at the rate of one leaf at the bottom of the baking pan for pound cake, venison, wild duck, roast beef, hamburgers, and chicken. Costmary tea is really a tonic and is still a favorite among herb teas — sometimes with mint.

Chapter 12

CURRY LEAF

I. *MURRAYA KOENIGII* (LINN) — FAMILY: RUTACEAE

Murraya koenigii (L.), or the Indian curry leaf is a small deciduous tree or shrub belonging to the natural order of Rutaceae. It generally grows in the wild and is a gregarious plant, often growing in tremendous profusion, particularly along the foothills of the Himalayas from Garhwal to Assam, up to an elevation of about 5000 ft in India. It is also quite widely distributed among the high forests of the Western ghats, extending southwards to the Karnataka, Kerala, and Tamil Nadu states in India. It is also found growing wild, although somewhat sparsely in the Vindhya hills embracing the Deccan Plateau in India. Mostly it is distributed in Konkan, Western ghats of Bombay to South India, and in most of the districts of Madras in South India and chiefly in the northern areas along the foot of the Himalayas from Kumaon to Sikkim up to 5000 ft in Bengal and Burma, and also in Sri Lanka.[3,12,62]

A. BOTANY[3,12,62]

Murraya koenigii Spreng is a small tree with dark gray bark. Leaves are imparipinnate and up to 30 cm long with petioles terete and pubescent. The leaflets are obliquely ovate or somewhat rhomboid, acuminate, obtuse or acute, with tip usually notched (the lower leaflet often suborbicular or obvate, much smaller than the upper), irregularly crenate, dentate, glabrous, pubescent beneath, and sprinkled with black dots. Flowers while in branched terminal are preduncled corrymtose cymes; peduncles and pedicles are pubescent. Calyx are pubescent, lobes subacute and triangular. Petals are 6 mm long, linear-oblong, rounded at the apex, and the gland is dotted. The filaments are dilated at the base. Ovary is two-celled, ovules solitary (rarely two) in each cell. Fruit ovoid or subglobose, 6 to 10 mm diameter apiculate, rough with glands, black, and two-seeded.[8,12]

In the submountain regions of the Himalayas, the plant often attains greater dimensions in the plains of India or in the Deccan plateau. The cultivated plant is always a small shrub. South Indian people cultivate this plant in their backyard or tiny gardens, exclusively for the sake of its aromatic leaves with a pronounced odor and flavor which appear to have a strong appeal to their culinary tastes. Northern Indians, however, seem to have less exposure to the culinary properties of the leaf.[112]

Propagation is by seeds, which germinate freely under partial shade; a spacing of 15 to 20 ft has been found optimum for planting seedlings. Many a time, dispersal is through birds which carry the red seed for a long distance.

B. COMPOSITION OF LEAF

The composition of *Murraya koenigii* is given in Table 17.

The leaves are a fair source of calcium but due to the presence of oxalic acid and oxalates (total 1.35%) and sodium oxalate (1.15%) in high concentration, its nutritional availability is affected. The leaves also contain a crystalline glucoside ("koenigin") and a resin. Twigs and leaves contain 0.8% potash (dry matter basis).

C. FREE AMINO ACIDS[13]

The free amino acids present in the leaves are threonine; alanine; tryptophan; phenylalnine; leucine; isoleucine; lysine; arginine; histidine; asparagine; glycine; serine; aspartic acid; glutamic acid; prolamine; γ-amino butyric acid, and traces of ornithine and tyrosine.

The chemical composition of green (fresh) curry leaves at three stages of maturity namely

TABLE 17
Composition of Fresh Curry Leaf (*Murraya koenigii*)[3,12,18,71,94,112,114]

Component	Content
Moisture (%)	66.3—68
Caloric value (kcal)	970
Protein (%)	6.1—7.0
Fat (%)	1.0—1.3
Carbohydrate (%)	16.0—18
Fiber (%)	5.5—6.4
Mineral (%)	4.0—4.2
Calcium (mg)	810—830
Phosphorus (mg)	50—60
Iron (mg)	3.0—7.0
Vitamin A (as β-carotene)	8,000—12,000 I.U.
Nicotinic acid (mg)	2.3—2.5
Vitamin C (mg)	4
Vitamin B_1 (mg)	80
Thiamine (mg)	1—5
Riboflavin (mg %)	208
Total oxalates (%)	1.35
Sodium oxalate (%)	1.15

TABLE 18
Composition of Different Stages of Curry Leaf on a Moisture Free Basis[3,117,118]

	Percent		
Content	Tender	Medium mature	Mature
Protein	5.4	6.4	7.2
Fat	3.3	4.7	6.2
Sugars	4.9	17.9	18.9
Starch	11.4	14.2	14.6
Crude fiber	5.8	6.2	6.2
Volatile oils	0.82	0.55	0.48
Acetone extract	6	1.4	3
Ash	12.5	12.7	13.1
Acid insoluble ash	1.2	1.3	1.4

tender, medium, and mature leaves was determined by Prakash and Natarjan[3,117,118] and is given in Table 18.

As can be seen from Table 18 with advancing maturity there is a gradual decrease in volatile oil and oleoresin while there is a progressive increase in many other contents.

Vacuum shelf drying of the curry leaves gave a better product of greenish color than those dried by other methods.[3,117,118]

D. ESSENTIAL OIL[56,113]

The essential oil was first isolated and examined by Guenther[56] who reported that the yield of the oil by ordinary steam distillation was about 0.04% and described the oil as a bright yellow mobile liquid. Dutt[113] distilled the fresh leaves of the plant obtained from various places in India for essential oil. The leaves obtained from central and western India had an extremely poor yield of oil of less than 0.2%. The aqueous distillate came over as a pale milky emulsion from which the oil could only be extracted by petroleum ether. During the course of the distillation, sulfurous (mercaptanic) compounds were detected in plenty.

TABLE 19
Distillation of Curry Leaf Oil[113]

Method employed	Yield of essential oil (%)	Specific gravity at 25°C	Refractive index at 25°C	Remarks
Simple water distillation	0.03	0.8814	1.4795	Yellow colored, extracted from aqueous distillate by petroleum ether.
With saturated steam (100—102°C)	0.05	0.8822	1.4832	Yellow colored, extracted from aqueous distillate by petroleum ether.
With steam at 90 lb psi (initial steam temp. 146—148°C)	2.6	0.9758	1.5026	Brownish-yellow colored, separates spontaneously from aq. distillate and rises to the top.
With saturated steam superheated to 220°C	2.9	0.9824	1.5086	Brownish-yellow colored, separates spontaneously from aq. distillate and slowly sinks to the bottom.

The essential oil contained in the leaves of *Murraya koenigii* is a comparatively heavy essential oil which is only slightly volatile in ordinary saturated steam. Hence, the idea of distillation with steam under pressure or with super heated steam was thought over by various workers. The fresh leaves with steam at 90 lb psi gave a brownish yellow oil. Table 19 gives the results obtained by steam distillation.

1. Distillation of Oil[116]

Nigam and Purohit[116] distilled the oil from water distillation of fresh leaves. The oil separated from the aqueous distillate as a light yellow layer floating over water which was carefully separated from water. The oil (1% yield) thus obtained was dried over anhydrous sodium sulfate. It gave a characteristic spicy smell and had the following constituents: specific gravity at 25°C: 0.8569; refractive index at 25°C: 1.4706; optical rotation in $CHCl_3$ at 25°C: 36°14′; acid value 1.17; saponification value 8.18; and saponification value after acetylation 52.81.

2. Fractionation of Oil[115,116]

The oil subjected to fractional distillation under reduced pressure (4 mmHg) yielded 12 different fractions.

Table 20 gives the fraction number and subfraction number with reference to boiling range, specific gravity, refractive index, and weight. Dutt[113] fractionally distilled the oil that was obtained at 90 lb psi with initial steaming between 146 to 148°C, (yield 2.6%) under reduced pressure and resolved into a light fraction (29%) consisting of mono- and bicyclic terpenes, and a heavy fraction (71%) consisting mainly of sesquiterpenes together with small quantities of higher aliphatic acids.

On studying the properties of this essential oil it was concluded that it is better to use the oil (if produced commercially) as an odoriferous principle or fixative of a moderately heavy (spicy) type of soap perfume. The compounds identified are given in Table 21.

3. Chromatography of the Oil[115,117,118]

The analysis of the essential oil of *Murraya Koenigii* (obtained by water distillation from fresh leaves and fractionating under reduced pressure) was achieved by the application of adsorption chromatography. Testing the fractions with chromatostrips (silica + starch) for the number of components contained therein was also performed.

TABLE 20
Fractionation of *Murraya Koenigii* Leaf Oil[115,116]

Fraction no.	Boiling range (°C)	Specific gravity at 25°C	Refractive index at 25°C	Yield (%)
1	30—35	0.8440	1.4660	1.3
2	35	0.8435	1.4665	19.3
3	36—38	0.8416	1.4670	9.2
4	38—40	0.8408	1.4675	12.7
5	40—45	0.8406	1.4695	10.1
6	45—48	0.8402	1.4720	10.5
7	48—80	0.8918	1.4760	5.3
8	75—85	—	—	2.5
	85—102	—	—	1.9
9	80—95	0.9112	1.4820	4.3
10	95—104	0.9158	1.5000	6.6
11	104—113	0.9200	1.5060	4.4
12	150—170	—	—	8.8
Residue	—	—	—	3.1

TABLE 21
Compounds Identified in the Fractions and Their Contents of *Murraya koenigii* Volatile Oil[113]

Fraction number	Compounds identified	Properties: Specific gravity (25°C)	Properties: Refractive index (25°C)	Properties: Optical rotation
1, 2	*l*-α-Pinene	0.8440	1.4665	−47°13′ in $CHCl_3$
3	Mixture of pinene and sabinene	—	—	—
4, 5, 6	*l*-Sabinene	0.8406	1.4675	−46°35′ in $CHCl_3$
7	Dipentene	0.8402	1.4715	±0 in $CHCl_3$
8	Difficult to identify since it was mixed with the 8th fraction	Similar to D-pentene, gave a crystalline tetrabromide, m.p. 124°C		
9	*l*-α-Terpineol	0.9296	1.4805	−21°28′
10	*l*-α-Caryophyllene	0.8913	1.5005	−5°26′
11	*l*-Cadinene	0.9137	1.5105	−80°21′

Preliminary chromatography (with alumina column) separated the oil into two portions, the hydrocarbons eluted with petroleum ether and the oxygenated compounds (α-terpineol, eluted with ethylacetate followed by alcohol). The separation of hydrocarbons was partial and it had to be completed by the repeated passage of the mixed fractions through fresh columns.

The advantage of this technique over the conventional methods of fractional distillation is that the possibility of polymerization and molecular rearrangement of the components are minimized. But a preliminary partial resolution of the mixture is necessary since one component of the oil may influence the adsorption of other.

By the above chromatographic separation of different terpenes, a relationship was established between their boiling point, structure of the molecule, the number of double bonds, and effect of various groups present in the molecule. The results are summarized in Table 22.

Unlike the essential oils obtained by previous workers, this oil was found to have a

TABLE 22
The Characteristics and Boiling Points of Some Components of the Oil of *Murraya Koenigii* Leaf

Compound	Boiling point (°C)	Structure of molecule	Number of double bonds	Absorption
Pinene	156—157	Bicyclic	One	Least
Sabinene	163—164	Bicyclic	One	Lesser
Dipentane	178	Monocyclic	Two	Less
Cadinene	274	Bicyclic	Two	More
L-caryophyllene	219—221	Monocyclic	Three	More
Terpineol	219—221	Monocyclic	One	Most

higher optical rotation. The oil consists of a greater percentage of simple monocyclic and bicyclic terpenes and a lesser percentage of sesquiterpenes as compared with the essential oil obtained by using steam under pressure. Moreover, no phenols and acids were detected when the oil was isolated under mild conditions of water distillation. A comparison of the physical properties showed that the drastic conditions though increasing the yield, spoil the essential oil as is evident from the darkening of its color and increase in its acid value. Caryophylene is the major constituent.

Table 23 gives a comparison of the properties of the oil of curry leaf as worked out by different workers.

4. Chemical Composition of the Oil

The oil (yield 2.6%) on chemical examination was found to contain the following constituents as shown in Tables 21 to 24. The physicochemical properties of the oil are specific gravity 0.9748 to 0.9824; optical rotation +4.8; acid value 3.8; ester value 5.2; and ester value after acetylation; 54.6. Nigam and Purohit[115] obtained, by water distillation of the fresh leaves, a yield of 1% oil as a bright yellow mobile liquid.

E. USES[3,6,12,71,74,112,113,117]

The leaves of *Murraya koenigii* (Linn) commonly known as curry leaf are extensively employed as flavoring in curries, chutneys, and soups especially in South India. It is an integral part of *rasam* in India without which the dish is not complete. The leaves are used whole and fresh and are fried for adding to spicy snack items. Leaves, bark, and root are tonic, stomatic, stimulant, and carminative. An infusion of the roasted leaves is used to stop vomiting. Along with *Mentha arvensis* in the form of chutney it is claimed to check vomiting. The green tender leaves are eaten raw for the cure of dysentery. When boiled in milk and ground they form a good application to poisonous bites and eruptions externally. Decoction of the leaves is given as a febrifuge in fevers. Leaves are also used as an antiperiodic and many a time the dry leaf, mixed with honey and juice of betel nut, is recommended in the ayurvedic system of medicine in India for better health.

II. *MURRAYA PANICULATA* (LINN)[62]

A. BOTANY[62]

This variety of curry leaf is an evergreen shrub or a small tree with spreading crown and short, often crooked, trunk found almost throughout India and the Andaman Islands up to an altitude of 1500 m. Bark is pale yellowish brown, rather corky and fragrant; leaves imparpinate 10 to 18 cm long; and petioles are glabrous. The terminal leaflet is the largest: ovate, elliptic, obovate, cyrhomboid, usually acuminate with a notched tip, entirely glabrous

TABLE 23
Comparison of the Physicochemical Properties and Chemical Composition of *Murraya koenigii* as Worked by Different Authors[113,114,116]

Properties	Essential oil (Penfold and Simosen[114])	Essential oil in fresh leaves (Dutt[113])		Essential oil (Nigam & Purohit[116])
Starting material	Dried leaves			Fresh leaves of young plants
Type of distillation	Simple steam	Steam distillation at 90 lb psi	Sat. steam super-heated to 20°C	Simple water distillation
Yield	0.04%	2.6%	2.9%	1.0%
Color	Bright yellow	Brownish yellow	Dark brown	Light yellow
Smell	Slightly disagreeable	Aromatic with strong spicy note	—	Aromatic with spicy note
Specific gravity	0.8711 (30°C)	0.9748(25°C)	0.9824(25°C)	0.8589(25°C)
Optical rotation	−28.2°	+4.8°	—	−36°14′
Acid value	1.1	3.8	—	1.17
Saponification value	11.06	5.2	—	8.18
Saponification value after acetylation	31.83	54.6	—	52.81
Chemical constituents	*dl*-sabinene	*dl*-Phellandrene (4.6%)		*l*-Pinene (27.3%)
	dl-Pinene	*d*-Sabinene	*l*-Sabinene (34%)	
	Dipentene	*d*-α-Pinene (5.5%)		Dipentene (15.9%)
	l-Caryophylene	Dipentene (6.8%)		*l*-Terpineol (7.7%)
	Unidentified	*d*-α-Terpineol (3.2%)		*l*-Caryophylene (6.7%)
	Sesquiterpene	Iso safrole (4.4%)		*l*-Cadinene (5.2%)
	Alcohol	Caryophylene (26.3%)		
	Palmitic acid	Cadinene (18.2%)		Unidentified residue 3.2%
	Unidentified	Cadinol (12.8%)		
	Volatile acid	Lauric acid (2.7%)		
		Palmitic acid (2.4%)		
		Residue (2.9%)		

TABLE 24
Chemical Composition of the Oil of Curry Leaves[3,74,113,117,118]

Constituent	Percentage
dl-α-Phellandrene	4.6
d-α-Pinene	5.5
β-Pinene	1.2
d-Sabinene	9.2
Dipentene	6.8
Isosafrole	4.4
d-α-Terpineol	3.2
Caryophyllene	26.3
Cadinene	18.0
Cadinol	12.8
Lauric acid	2.7
Palmitic acid	2.4
Residue (unidentified)	2.9

and shining; the base is acute and oblique; and petioles are 3 to 6 mm long. Flowers are very fragrant, companulate, solitary, or in terminal and axillary corymbs; septals are small, glandular, oblong, and obtuse; petals are white, 1.3 to 2 cm long, oblong, lanceolate, subobtuse, erect at the base, with the upper half spreading. Filaments are flat, linear, and tapering beneath the anther. The ovary is two-celled; the berry is 1.3 to 2 cm diam; oblong, pointed, smooth, and one-celled, two-seeded, and red when ripe. The plant is commonly grown in gardens for its glossy green foliage and large clusters of fragrant flowers. It is a popular hedge plant and is well adapted for topiary work. Propagation may be done by seeds, cuttings, or layering. The plant is subject to the attack of the citrus stem borer, *Chelidonium cinctum* guer. Removal of infected branches or treatment of affected parts with chloroform and creosote mixture is recommended as control measures.[3]

B. VOLATILE OIL[3,114]

Fresh leaves of *Murraya paniculata* on steam distillation yield (0.01%) a dark-colored volatile oil with a pleasant odor.

Mehrotra and Gupta[114] extracted the oil from fresh leaves by steam distillation and the distillate was treated with purified petroleum ether (40 to 60°C) to remove the essential oil. Petroleum ether was later removed by distillation leaving behind the oil. The yield of the oil was 0.04% on the weight of fresh leaves. The properties of the oil are reported by Mehrotra and Gupta.[114]

C. CHEMICAL COMPOSITION[114]

The oil contains sesquiterpenes (predominantly *l*-cadinene), a sesquiterpene alcohol and probably methyl anthranilate.

Flowers are reported to contain indole and a bitter crystalline glucoside murrayin (yield 1.3%), which is identical with saponin (7-glucoside-6-methoxycoumarin, $C_{16}H_{18}O_9$, m.p. 218°C).

D. USES[3,12,62,71,74]

The leaves are stimulant and astringent and are administered in some of the southeast Asian countries in the treatment of diarrhea and dysentery. They are given in the form of an infusion in the proportion of 15 g/liter of water.

Powdered leaves are applied to cuts. Leaves and root bark are sometimes used against rheumatism and other nervous disorders. Twigs are used for cleaning teeth. The leaves are reported to possess antimicrobial activity against *Micrococcus pyogens* var. *aureus* and *Escherichia coli*. The decoction of the leaves is drunk in dropsy. In India the ground bark of the tree is used as a drink in snake bites and rubbed on the bitten limb by certain tribes. The ground bark of the root is rubbed for body ache. Leaves and root bark are sometimes used against rheumatism and cough.

Chapter 13

MARJORAM

I. *MAJORANA HORTENSIS* (M.) — FAMILY: LABIATAE

For many years both marjoram and oregano were known as *Origanum majorana* L. Today Marjoram is identified as *Majorana hortensis* as a member of the mint family indigenous to the Mediterranean region. It is believed to have originated from Arabia but today one finds it being cultivated in France, Germany, Great Britain, Italy, Morocco, Spain and the U.S.[3]

A. BOTANY AND CULTIVATION[1,3,58,71,74]

Sweet or knotted marjoram is a low, tender perennial 12 to 18 in. high, which is grown as an annual in the northern climates. Its downy, narrow leaves are light gray-green in color, about 1/2 in. long. The inconspicuous white or pinkish flowers develop in terminal clusters. The plant has a pleasant aromatic aroma. The flavor is warm, aromatic, slightly sharp, bitter, and camphoraceous. Sweet marjoram has a delicate, pleasant, sweet flavor. Marjoram is planted in rows and intervals of about 9 × 9 in., oregano about 12 × 12 in. apart.

The first harvest of the leaves and tender tops of both herbs occurs as flowering commences, usually during July or August. The plants are cut 2 to 3 in. from the ground, and with favorable conditions a second cutting may be made in October. The oregano planting may be productive for 4 to 5 years, but in North Europe marjoram is usually renewed annually. After the harvest the leaves are thoroughly dried, carefully cleaned, and stored.

The chief structural characteristics of the leaf are that it is spatulate, petioled, and downy haired on both sides. The hairs are pointed, long, broad, joined smooth and warty, and stiff only on petioles, ribs, and veins. They are also capitate with one to two joints. Bladder glands have 8 to 12 cells.[3,17] In the case of *Origanum vulgare* L. the hairs are up to 1 mm long.

Methods of drying depend upon the size of the crop and the climatic conditions in the producing countries. The cut plants may be tied in bunches and dried in open air or spread on wire trays in ventilated rooms and dried by controlling the circulating warm air. The aroma of dried marjoram is fragrant, the taste aromatic, warm, slightly sharp, and somewhat bitter. It is available whole or ground in the market.

B. SPECIFICATIONS IN U.S.[5]

Marjoram shall be the dried leaves, with or without a small portion of the flowering tops of *Majorana hortensis* Moench. The round, light green to light gray-green leaves should possess a pleasant aromatic odor and have a warm, slightly bitter taste. The ground product shall contain not more than 10% of stems by weight; and it must be uniformly ground to allow for a minimum of 95% by weight, to pass through a U.S. standard no. 30 sieve. The product shall contain not more than 10% moisture, 13% total ash, and 4.0% acid insoluble ash. It shall contain not less than 0.8% volatile oil expressed as ml per 100 g.

The composition of the dried leaves is given in Table 25.

C. EXTRACTIVES[5]

The essential oil is obtained by steam distillation. The yield is very low (less than 1.0%), it is a yellow or greenish-yellow extract with an aroma that is spicy, fragrant, warm, aromatic, penetrating, and resembling that of lavender. The taste is sharp, warm, spicy, slightly bitter, herbaceous, and smoothly aromatic with a slightly bitter aftertaste.

The oleoresin is a dark green, viscous, semisolid, with a volatile oil content of 40 ml/

TABLE 25
Composition of Sweet Marjoram (100 g Edible Portion)[5]

Constituent	Quantity
Water	7.6 g
Food energy	271 kcal
Protein	12.7—14.3 g
Fat	5.6—7.0 g
Total carbohydrate	60.6 g
Pentosans (% carbohydrates)	7.68
Fiber (% carbohydrates)	18—22 g
Ash	12—13.0 g
Calcium	1990 mg
Iron	83 mg
Magnesium	346 mg
Phosphorus	306 mg
Potassium	1522 mg
Sodium	77 mg
Zinc	4 mg
Ascorbic acid	51 mg
Niacin	4 mg
Vitamin A (as β-carotene)	8068 I.U.

TABLE 26
The Mineral Constituents of German and French Varieties of Marjoram[17]

Minerals as oxides	German oil (%)	French oil (%)
K_2O	20.18	18.34
Na_2O	0.68	0.65
CaO	16.70	24.80
MgO	4.76	6.74
Fe_2O_3	7.30	6.06
Mn_3O_4	1.05	Trace
P_2O_5	8.88	9.10
SO_3	4.92	4.80
SiO_2	26.52	19.44
Cl	2.05	1.51
CO_2	6.06	8.56

100 g. For a good quality oleoresin, 1.14 kg is equivalent to 45.45 kg of freshly ground marjoram in odor and flavor characteristics.

D. MINERAL CONSTITUENTS[3,17]

Wide variation in the mineral content of Indian, French, and German sweet marjoram herb is reported. Total ash 6.3 to 24%; sand 0.66 to 14%; sand-free ash 5.4 to 14.3%; potash 18.3 to 20.2%; sodium 0.65 to 0.68%; calcium 17.6 to 24.8%; phosphorus 8.9 to 9.1%; iron 6.1 to 6.3%; silica 19.4 to 26.5%; magnesium 4.8 to 6.7%; manganese trace to 1.05%; and chlorine 1.51 to 2.05%.[3,6] The wide range of mineral constituents of various varieties include 6.3 to 24.0% in ash; 0.66 to 14.0% in sand; and 5.4 to 14.3% in sand-free ash. The average ash content of 156 samples and the respective mineral oxide are shown in Table 26.

TABLE 27
The Physicochemical Properties of the Oil of *Origanum majorana*[55,56]

Constants	U.S.	Italy	India	Hungary
Specific gravity at 15°C	0.892—0.9164	0.8840—0.8960	0.9346	0.9099—0.9164
Optical rotation	+13°25′—32°24′	+12°0′ +15°24′	+40°15′	+13°25—′ +32°24′
Refractive index at 20°C	1.4653—1.4821	1.4730—1.4805	1.5062	1.4653—1.4821
Acid number	1.4—2.8	—	4.8	—
Saponification number	32.7—40.1	8.5—19.96	8.32	—
Ester content	—	2.77—6.24	—	—
Ester number after acetylation	30.94—86.8	—	—	38.94—60.07
Saponification number after acetylation	—	—	128.4	—
Phenol content	—	—	47.7%	—
Alcohol content	—	32.4—41.31%	—	—
Solubility	1—2 vols of 80% alc.	0.8—1.2 vols of 85% alc.	—	20—40 vols of 70% alc.

E. VOLATILE OIL[55,56,119]

A yield of 0.3 to 0.4% for fresh, flavoring herbs and a yield of 0.7 to 3.5% for the dried herb material is reported. The physicochemical properties of the oil vary according to each author and a summary is presented in Table 27. Lawrence[120] has summarized the work of nearly 12 authors during the period of 1961 to 1982 along with his own analysis of the oil.

The yellowish oil has a pleasant aromatic odor. The surge of information on volatile oil and its constituents was around 1960. Since then newer components have been identified using modern techniques.

Khanna et al.[119] have reported the following constituents of the oil; α-pinene (5.6%), α-thujone (4.1%), camphene (0.4%), α-terpinene (6 to 18%), myrcene (7.2%), *d*-limonene (1.0%), β-phellandrene (0.6%), γ-terpinene (35.1%), *p*-cymene (39.3%), thymol (1.6%), linalool (3.7%), geraniol (3.3%), and carvacrol (4.9%).

Nicoletti and Baiocchi[121] have identified terpinene-4-ol (35.5%), sabinene hydrate (9%), linalyl acetate (8%), linalool (4.5%), and α-terpineol (4.5%), in an oil of marjoram. The following year Ikeda et al.[79] and Dayal and Purohit[122] used retention time data to characterize the hydrocarbons (36.4% of the oil) found in sweet marjoram oil. Using this procedure, the authors identified α-pinene (3.0%), α-thujone (3.1%), camphene (0.2%), β-pinene (1.1%), sabinene (11.0%), α-phellandrene (1.8%), α-terpinene (18.1%), myrcene (5.5%), limonene (4.9%), β-phellandrene (5.0%), γ-terpinene (28.1%), *p*-cymene and terpinolene (17.2%) and an ocimene isomer (1.0%).

Vashist et al.[123] reported that marjoram oil obtained from European plant material raised in North India contained α-terpinene (24.1%), terpinen-4-ol (10.2%), α-terpineol (26.7%), linalool (5.6%), geraniol (19.0%), and eugenol (8.0%). Later Lossner[124] found that French marjoram leaves possessed an oil which contained *cis*-sabinene hydrate (25%) and *trans*-sabinene hydrate (7%).

Karawya et al.[125] used a combination of thin-layer chromatography, column chromatography and gas chromatography to characterize the presence of α-pinene (2.5%), camphene (2.8%), β-pinene (0.2%), sabinene (0.9%), myrcene (0.3%), limonene (5.5%), β-phellandrene (31.9%), ocimeneno isomer (13.9%), 1,8-cineole (11.7%), camphor (5.6%), linalool (10.6%), linalyl acetate (2.0%), bornyl acetate (0.1%), borneol (0.5%), and α-terpineol (trace) in Egyptian marjoram oil. Rey and Boussaire[126] reported that the Italian marjoram oil contained a mixture of thymol (major component) and carvacrol (46.7% total phenols).

Granger et al.[127] examined the chemical composition of an Egyptian oil of marjoram and compared it to a French oil. The oils were found to contain α-pinene (2.3%) (1.5%); β-pinene (0.2%) (02%); sabinene (8.1%) (9.5%); myrcene (3.2%) (2.5%); α-terpinene (10.8%) (6.2%); limonene (2.0%) (2.0%); β-phellandrene (3.6%) (1.8%); γ-terpinene (17.0%) (12.2%); *p*-cymene (5.5%) (1.5%); terpinolene (4.4%) (2.6%); *trans*-sabinene hydrate (1.8%) (5.5%); *cis*-sabinene hydrate (0.5%) (17.0%); linalool (2.5%) (3.0%); linalyl acetate (2.8%) (3.5%); terpinen-4-ol (20.0%) (21.2%); caryophyllene (3.2%) (1.0%); α-humulene (0.5%) (0.3%); and α-terpineol + α-terpinyl acetate (4.3%) (2.2%), respectively. In addition, the same authors examined six other French (Drome) marjoram oils. In these they identified α-pinene (1.3 to 2.0%); β-pinene (0.1 to 0.2%); sabinene (9.0 to 11.9%); myrcene (2.5 to 3.6%); α-terpinene (0.8 to 1.9%); limonene (0.8 to 2.3%); β-phellandrene (0.6 to 2.3%); γ-terpinene (3.2 to 4.2%); *p*-cymene (0.3 to 1.4%); terpinolene (0.2 to 0.9%); *trans*-sabinene hydrate (6.8 to 8.1%); *cis*-sabinene hydrate (34.0 to 37.2%); linalool (1.5 to 2.0%); linalyl acetate (2.0 to 3.5%); terpinen-4-ol (4.3 to 7.5%); caryophyllene (2.2 to 3.6%); α-humulene (0.1%); and α-terpineol + α-terpinyl acetate (2.3 to 2.8%). The authors also isolated another compound (12.4—15.0%) in these oils which was not identified. The authors showed that *trans*-sabinene hydrate could be readily hydrolyzed to terpinen-4-ol and monoterpene hydrocarbons depending upon pH, temperature, and available water. This has a bearing on the hydrolysis of *trans*-sabinene hydrate which can easily take place during distillation of the oil especially if it is water distilled.

Masada[128] reported that α-pinene, camphene, β-pinene, α-terpinene, 1,8-cineole, γ-terpinene, *p*-cymene, camphor, linalool (major constituent), methyl chavicol, α-terpineol, and eugenol were identified in an oil of marjoram obtained from cultivated plants. Croteau[129] examined the site of monoterpene biosynthesis in marjoram leaves and showed that α-terpinene, γ-terpinene, *cis*-sabinene hydrate, linalool, *trans*-sabinene hydrate, terpinen-4-ol, and α-terpineol were the major monoterpenoid constituents in the leaves.

Vernon et al.[130] have identified the following components in the French and Egyptian oils respectively: α-thujene + α-pinene (1.45%, 1.19%), sabinene (5.69%, 4.54%), myrcene (1.02%, 0.07%), δ 3-carene (0.11%, 0.02%), α-terpinene (1.13%, 0.10%), α-phellandrene (—, 0.93%), limonene + β-phellandrene (7.78%, 0.13%), *p*-cymene (11.90%, 27.26%), terpinolene (1.75%, 0.04%), *trans*-sabinene hydrate (1.67%, 2.61%), linalool (1.81%, 2.06%), *cis*-*p*-menth-2-en-1-ol (—, 0.58%), *cis*-sabinene hydrate (8.91%, 7.14%), linalyl acetate (0.51%, 1.51%), terpinen-4-ol (36.33%, 35.10%), caryophellene (2.44%, 2.46%), α-terpineol (4.30%, 4.86%), geraniol (0.54%, 1.83%), geranyl acetate (—, 0.15%), and *p*-cymen-8-ol (—, 0.64%).

Rhyu[131] examined the oils obtained from laboratory distillations of marjoram leaves obtained from Chile, Egypt, and the U.S. and reported nearly 22 components.

The following are the percent composition for the above three varieties of oil respectively as shown in parentheses (average): α-pinene (1.15, 0.4, 0.2), camphene (0.1, 0.1, trace), β-pinene (0.4, 0.15, trace), sabinene (3.45, 3.0, 0.4), myrcene (1.1, 1.5, trace), α-phellandrene (0.7, 1.15, 0.1), α-terpinene (9.25, 8.25, 7.2), limonene (1.4, 1.15, 0.3), 1,8-cineole (5.05, 1.85, 0.6), γ-terpinene (11.7, 14.3, 3.0), *p*-cymene (2.5, 2.8, 0.7), terpinolene (2.6, 2.5, 0.7), thujene (traces), camphor (traces), copaene (0.4, trace, 0.5), linalool (1.8, 1.25, 0.3), terpinene-4-ol (19.5, 26.0, 2.8), caryophyllene (3.05, 2.25, 1.3), methyl chavicol (traces), α-terpineol (1.9, 3.75, 0.3), thymol (traces, 12.4, trace), carvacrol (trace, 2.2, trace) and unknowns (15, 2, 35). However, this work did not represent the full profile of the oil since some of the major established constituents were not characterized.

Subsequently, Oberdieck[132] compared the chemical composition of oils obtained from commercial samples of marjoram leaves. Using a combination of GC, IR, and MS, the composition of two German oils, two Hungarian oils, and oil of Romanian, Portuguese, Egyptian, and Tunisian origins have been compared and tabulated with exhaustive information.[132] Later, Brosche et al.[133] used ^{13}C-NMR, GC, and GC-MS to examine the composition of marjoram oil. In this study the authors identified α-thujene (0.35%), α-pinene (0.4%), sabinene (5.6%), β-pinene (0.2%), myrcene (1.5%), α-phellandrene (0.2%), α-terpinene (7.3%), *p*-cymene (0.1%), limonene + β-phellandrene (2.6%), γ-terpinene (11.4%), *trans*-sabinene hydrate (4.25%), terpinolene (2.3%), *cis*-sabinene hydrate (21.1%), linalool (2.5%), *trans-p*-menth-2-2n-1-ol- (1.5%), *cis-p*-menth-2-en-1-ol (0.7%), terpinene-4-ol (23.5%), α-terpineol (3.7%), trans-piperitol (0.4%), *cis*-piperitol (0.4%), *cis*-sabinene hydrate acetate (0.3%), linalyl acetate (3.3%), terpinen-4-yl acetate (0.2%), neryl acetate (0.1%), geranyl acetate (0.2%), caryophyllene (1.7%), α-humulene (0.1%), and γ-elemene (1.2%). Traces of xylene, nonane, camphene, 1,8-cineole, β-terpineol, isopulegol, thujol, *trans*-sabinene hydrate acetate, nerol, geranil, bornyl acetate, α-terpineyl acetate, aromadendrene, allaromadendrene and δ-cadinene were identified.[120,133,137]

Lawrence[120,134,136] also examined the oil of marjoram and a summary of this analysis of laboratory distilled oil includes: α-thujene (0.2%), α-pinene (0.6%), β-pinene (0.2%), sabinene (2.8%), myrcene (0.8%), α-phellandrene (0.3%), α-terpinene (3.4%), limonene (0.8%), β-phellandrene (0.9%), 1,8-cineole (0.1%), *cis*-ocimene (0.1%), α-terpinene + *trans*-ocimene (7.6%), *p*-cymene (3.2%), terpinolene (1.1%), α-*p*-dimethyl styrene (0.1%), *trans*-sabinene hydrate (0.7%), linalool (1.0%), *cis*-sabinene hydrate (7.8%), linalyl acetate (2.2%), *cis-p*-menth-2-en-1-ol (2.4%), terpinen-4-ol (46.1%), terpinen-4-yl acetate (0.2%), caryophellene (2.6%), *trans-p*-menth-2-en-1-ol (1.6%), α-humulene (0.2%), α-terpineol (7.6%), α-terpinyl acetate (0.2%), neryl acetate (0.4%), bicyclogermacene (1.1%), geranyl acetate (0.6%), *p*-cymen-8-ol (0.1%), caryophyllene oxide (0.1%), and elemol (0.2%).

Recently Sarer et al.[135] examined the chemical composition of *Origanum majorana* oil which was obtained from plants found growing wild in Turkey. Using column chromatography as the method of prefractionation, the authors found that the oil contained monoterpene hydrocarbons (12.6%), oxygenated monoterpenes (18.4%), and phenols (69.0%). Analysis of the various fractions using retention times and peak enrichment on three different columns revealed that the following prefractions were found to contain a number of constituents. Monoterpene hydrocarbons: tricyclene(t)α-pinene + α-thujene (9.2%), camphene (1.7%), β-pinene (1.0%), δ-4-carene (0.1%), δ-3-carene (0.4%), myrcene + α-phellandrene (10.6%), α-terpinene (7.4%), limonene (1.4%), β-phellandrene (2.9%), *cis*-ocimene (1.0%), γ-terpinene (24.8%), *trans*-ocimene (0.5%), *p*-cymene (36.0%), and terpinolene (0.7%). Oxygenated monoterpenes: 1,8-cineole (1.4%), 3-octanol (0.3%), fenchone (1.5%), bornyl acetate (1.0%), terpinene-4-ol (3.6%), *trans-p*-menth-2-en-1-ol (0.2%), *trans*-sabinene hydrate (2.7%), β-thujone (0.3%), menthone(t)linalool (76.6%), *cis*-sabinene hydrate (1.6%), *cis-p*-menth-2-en-1-ol (0.5%), isoborneol (1.0%), α-terpineol (3.3%), borneol (2.8%), carvone (0.6%), geraniol (0.2%). Phenols: carvacrol (65.1%) and thymol (3.9%).

F. USES[1-6,16,60,70,74,78]

The dried or fresh leaves are used to flavor innumerable varieties of food and marjoram is one of the most useful of all culinary seasonings. The young tender leaves fresh or dried, may be used with seafood, soups, eggs, fish, meats, poultry, game, stews, vegetables, salads, and sauces. The powdered marjoram is often obtained with other herbs in preparing specific herb blends. It is always an ingredient in poultry seasoning and is used extensively in sausage manufacture. One teaspoon of ground marjoram equals 0.051 ounce.[70] It may be used sparingly with highly spiced foods like pizzas and many Italian, Greek, and French recipes.

Marjoram is carminative, stimulant, diaphoretic, emmenagogue, and tonic. Volatile oil is used as an aromatic stimulant in colic, dyspepsia, flatulence, and dysmenorhoea; the dose is 2 to 5 minims. Like *Oleum mentha* it is used locally in rheumatism, to abdomen in colic, to the temples in hemicrania and to the ear in earache. Infusion of the plant (1 in 10) is also useful for internal administration in doses of half to one ounce and externally for fomentation.[121]

II. *ORIGANUM ONITES* (POT MARJORAM)

Origanum onites is frequently grown because sweet marjoram is not hardy, although it does not have as sweet a flavor and is even slightly bitter. This is also a Mediterranean plant. It grows about 1 ft high and the flowers are white. Pot marjoram can be used to some extent for the same purposes as sweet marjoram especially in the more strongly flavored dishes such as with onion, wine, and garlic, where the delicate perfume of sweet marjoram would, in any case, be largely lost. It grows wild in Greece and is one of the plants they call *rigani*.[3]

There are nearly 10 different wild species of origanum known commonly under the name of *rigani* in Greece. One of these known as winter marjoram (*Origanum heraclesticum*) is sometimes cultivated in gardens. Other species are *Origanum smyrnaicum* and *Origanum paniflorum*. Rigani is used with grilled meats and other Greek dishes, but it is almost impossible at present to buy the authentic herbs and exactly reproduce the flavor of such dishes outside the country. Another famous species of *Origanum* commonly known as ''cretan dittany'' is cultivated particularly on the island of Crete and is used mainly for medicinal purposes and also as a food flavoring.[3]

Chapter 14

MINT — FAMILY: LABIATAE

Mint belongs to a small genus of aromatic perennial herbs distributed mainly in the temperate regions of the world. Several species have been introduced into various countries and are cultivated for their aromatic leaves and flowers. About six species are recorded in India where it is used profusely.

Mentha species are extremely variable and specific, limits are hard to define, as a consequence, unstable nomenclature is in vogue. All cultivated types are said to belong to three or four species which have hybridized among themselves, resulting in a number of inter-grading hybrids, often misgraded as separate species. They appear to differ not only in morphological features, but also in flavor and essential oil content. Many of them have escaped cultivation and grow naturally in moist conditions; some are sterile or only half fertile and have been perpetuated by vegetative propagation.[3,138]

Of the many species of mint, peppermint — *Mentha piperita* spearmint and *Mentha spicata* are among the common and foremost flavoring used in foods. They are indigenous to Europe and the Mediterranean region but naturalized throughout much of the temperate world in both Northern and Southern Hemispheres. Historically, spearmint is much older than peppermint. Commercial mint culture is extensively used today in the U.S. especially in California, Oregon, Washington, Michigan, Ohio, Wisconsin, and Indiana. Other countries important in the production of mint include Romania, England, France, Egypt, Argentina, Russia, Bulgaria, Morocco, and India.

Mint planting is productive for 5 to 6 years, hence, its site should be selected carefully. It should be rich, well-drained loam soil with adequate moisture and plenty of sunlight.

The distinction between the two species, peppermint and spearmint, may be classified as ('P') for peppermint which has a petioled (stalked) leaf while 'S' is for spearmint with a sessile (nonstalked) leaf.

The peppermint plant, somewhat taller and more reddish than spearmint, may grow to a height of 2 to 3 ft. Its leaves are bright, green in color, more clearly stalked, and more pungent in taste than spearmint. The flowers of both varieties are pale violet.

The white mint variety is grown on a limited scale in southern England. Because of its susceptibility to serious infections there is difficulty in growing it economically. It has been replaced with the hardier, black mint variety. It is believed that white mint has better flavor.[5]

The black mint variety grows to a height of 1 m (3 ft) and bears bright green leaves measuring up to 45 mm long by 23 mm wide. The plant is propagated by cuttings. The green leaves and tops are harvested when in full bloom and partially dried with the leaf stalks attached. The aroma is very strongly mentholic, sweet and cooling, with a taste that is fragrant, spicy, minty cool, and slightly pungent. The aftertaste is sweet, minty, and cooling.[5]

Punia et al.[138] conducted experiments on four species of mint; *Mentha arvensis, Mentha piperita, Mentha spicata* and *M. citrata*, followed by *Mentha arvensis*. The highest average oil content was found in the species of *Mentha arvensis* (0.61%) followed by *Mentha citrata* (0.42%) and *Mentha piperita* (0.34%); *Mentha spicata* contained the lowest oil content (0.26%).

I. *MENTHA PIPERITA* (LINN)

A. BOTANY[3,12,19,62,69,75,101,139]

Peppermint is native to Mediterranean countries. Commercially grown species of *Mentha piperita* L. have two recognized varieties: *M. pripetu* var. *vulgaris* commonly known as the

English or blackmint and *Mentha piperita* (L.) var. *officinalis* popularly called whitemint. The latter variety is not grown on a commercial scale owing to its low yield. However, the oil obtained from this variety enjoys a high reputation and commands a higher price than that of blackmint.

B. HARVESTING[14]

Peppermint should be harvested when the plant starts to bloom so as to ensure the maximum yield of the oil as well as menthol content. The oil decreases rapidly after full bloom stage of the plant due to shedding of the leaves in which oil is present. In case the harvest is late, the loss of the oil is greater than if the crop is cut prematurely.[101]

The oil content of the peppermint plant increases up to a certain maturity after which the yield of the oil decreases and one of the constituents of oil, free menthol increases. The plant first develops menthone which is transformed into menthol at a later stage. Hence the appearance of bloom is supposed to be an indication of the maximum level of oil.

C. DRYING AND DISTILLATION[3,6,13]

The leaves are spread under shade and turned over every morning until it is half dry. It takes about 4 to 5 days to dry it to a reasonable moisture level. It may be mentioned here that exposure of the fresh herbage to direct sunlight results in considerable loss of volatile oil either by evaporation or by resinification. The leaves of plants of labiate family in general and mint in particular when dried in direct sun, lose as much as 24% of essential oil as compared to leaves dried in the shade where the loss is between 2 to 10%. The dry herbage is more economical to handle and yields oil much more readily upon distillation with economic considerations.

D. VOLATILE OIL[3,13,14]

The physical and chemical constants of the oil of peppermint of different sources are tabulated in Table 28. The chief constituents of the oil are *l*-menthol (45 to 60%) and *l*-menthone (15 to 30%). Menthol is responsible for the cooling aftertaste.

The constituents of the essential oil as reported by different workers are[3,13,141] α-pinene (0.2%), β-pinene (0.9%), limonene (3.5%), cineole (4.0%), sabinene hydrate (0.6%), sabinene acetate, menthofuran (0.7%), isomenthane (5.0%), menthane (7.7%), neomenthol (4.1%), menthyl acetate (7.5%), menthol (60.6%), pulegone (1.5%), piperitone (2.0%), piperitone oxide (0.1%), acetaldehyde (0.044%), dimethyl sulfide, isovaleric aldehyde (0.048%), isoamyl alcohol (2 methyl 4-butanol amyl alcohol phellandrene) terpinine, *d*-menthone, menthanone, *p*-menthen-3-one, menthyl isovalerate, cadinene, lactone, menthofuran, octylic acid, thymol, carvacol, jasmone, l-caryophyllene, 3-methyl-δ-l-cyclohexane, dipentene, cadinene, and piperitone (mint glyoxal).

E. USES[1-6,12,13,52,62,74,75,142]

Essential oil of mint is widely used for flavoring of pharmaceuticals and oral preparations such as tooth pastes, dental creams, mouth washes, cough drops, chewing gums, confectionery, and alcoholic liquors. So far as the medicinal value is concerned the oil is generally preferred to menthol for internal use because of its more pleasant aroma and taste. Oil of peppermint is an excellent carminative and a gastric stimulant. In the alimentary canal, it aids as an antiseptic and has local anesthetic influence.

It is used as an antiseptic, deodorant, stimulant, and carminative. Externally it is applied in congestive headaches, rheumatism, and neuralgia. Pharmaceutical preparations are done with mint oil to disguise the taste of bad smelling and unpleasant drugs and as a flavoring in confectionery and dentrifices. Leaves are aromatic, stimulant, carminative, and antispasmodic. Leaves in infusion (1 in 10) or their oil or as a spirit in doses of 5 to 20 minims

TABLE 28
Physicochemical Properties and Important Chemical Constituents of the Oil of Peppermint from Different Sources[13,75,140]

	U.S.	U.S.S.R.	Bulgaria	Italy	France
Specific gravity at 15/15°	0.9140	0.899—0.909	0.907—0.9135	0.902—0.926	0.910—0.927
Optical rotation	−32°	−26°45′—−21°	−18°0′—22°12′	2°30′—−27°0′	−5°0′—−35°
Refractive index	1.454—1.464	1.459—1.472	1.461—1.464	1.462—1.470	1.462—1.471
Acid number	—	—	—	Up to 0.6	—
Total menthol	59.6%	47.0—57.0%	52.0—59.3	43.0—67%	45.0—70.0
Ester menthol	14.12%	3.1—12.5%	4.3—6.6	2.9—10.4	4.0—21.0%
Menthane content	—	21.0—25.4	—	8.0—29.4%	17.4%
Solubility in alcohol	—	3 vols of 70%	3 vols of 70%	3 and more vols of 70%	3.5 vols of 70%

	Argentina	Romania	Hungary	England
Specific gravity at 15/15°	0.905—0.902	0.9042—0.9113	0.8894—0.9135	0.901—0.912
Optical rotation	−24°10′—−22°58′	−21°10′—27°9′	−14°—−38°30′	−18°6′—−33°6′
Refractive index	—	1.4602—1.4618	1.4610—1.4769	1.4591—1.467
Acid number	—	—	—	Up to 1.6
Total menthol	65.2—58.3%	50.0—65.8%	36.2—71.37%	42.4—68%
Ester menthol	7.3—6.3%	3.3—12.5%	5.07—56.3%	2.3—21.0
Menthane content	19.0—19.6%	7.9—10.1%	7.40—14.11%	9—42%
Solubility in alcohol	2.5—3 vols of 70%	2.7 vols of 70%	—	3 vols of 70%

	Indian	I.S.I. specifications
Specific gravity	0.9046 (at 27°)	0.896—0.908 (at 30/30°)
Optical rotation	−30.3°	−18°—−36°
Refractive index	1.4632 (at 27°)	1.4555—1.4655 (at 30°)
Acid value	1.4	—
Menthol (%)	55.8 (free)	+50 (50)
Ester (as menthyl acetate)	20.00	4.20
Menthane (%)	9.1	25
Solubility in 70% alcohol	Soluble in 2.7 vol	Soluble in 3.5 vol

TABLE 29
Composition of *Mentha spicata* Linn[3,12,18]

Constituents	Quantity
Moisture	83.0—85.0%
Protein	4.8%
Fat (ether extr.)	0.6%
Carbohydrates	6.0—8.0%
Fiber	2.0%
Mineral matter	1.6—1.9%
Calcium	200 mg%
Phosphorus	80 mg%
Iron	15.6 mg
Carotene (as Vit A)	2,700 I.U.
Nicotinic acid	0.4 mg
Riboflavin	80—100 μg
Thiamine	30—50 μg/100 g
Copper	1.8 μg/g
Niacin	1.0 mg
Vitamin C	27 mg

of aqua in doses of 1/2 to 2 ounces are used in cases of vomiting, gastric colitis, cholera, diarrhea, flatulence, etc. It is given in dysmenorrhea, together with tea in weak digestion, in hiccups, and palpitation of the heart. It is given with purgatives as a corrective and preventative for gripping. Locally the oil is a powerful anodyne, anesthetic, antiseptic, and germicide useful in herpes zoster, pruritus, etc. in the form of lotion. It is used as a paint in diphtheria. It is used to relieve toothache caused by caries also.

II. *MENTHA SPICATA* (L.)[12]

A. BOTANY[3,12,13,75]

Mentha spicata is a glabrous perennial, 30 to 90 cm high, with creeping rhizomes indigenous to the north of England, but is also grown all over the world. It is cultivated in Indian gardens also. Leaves are smooth or nearly so, senile, lanceolate to ovate, acute, coarsely dentate, smooth above, glandular below; the flowers are lilac, loose, cylindrical, and slender, with interrupted spikes. Newer strains are in cultivation now.[143,154]

This species is variable and is often erroneously recorded under the name *M. viridis*. It includes a number of forms whose identity and nomenclature are confusing. The species itself is considered to be a hybrid between *M. rotundifolia* and *M. longifolia*. Cytological evidence indicates that the forms vary greatly in chromosome numbers and essential oil content.

Spearmint is widely cultivated throughout the plains of India and used for culinary purposes. The U.S. accounts for most of the world output of spearmint. It is cultivated extensively in Russia, Japan, Great Britain, Germany, and The Netherlands.

It thrives best in heavy loams well-supplied with farmyard manure. It is usually propagated by planting divisions of old plants in rows of 30 cm apart, the distance between the plants in the row being 15 cm. The field is weeded and watered during dry weather. The plants produce leaves for a number of years, but it is advisable to replant them annually in order to get tender leaves and luxuriant growth. The leaves have a characteristic odor and a slightly pungent taste, not followed by a cooling sensation as in the case of peppermint.

B. COMPOSITION

The composition of green leaves of *M. spicata* is presented in Table 29.

TABLE 30
Physicochemical Properties of Spearmint Oils from Different Parts of India[3,12,144-147,158]

Properties	Kanpur (plains)	Sub-Himalayan: Kashmir	Sub-Himalayan: Dehradun	Poona (Deccan plateau)
Specific gravity	0.9349 (at 30°)	0.94 (at 18°)	0.9482—0.9236 (at 20°)	0.9817 (at 30°)
Optical rotation	−57.4°	—	—	−97.8°
Refractive index	1.4948 (at 30°)	1.539 (at 18°)	1.4880—1.4900	1.4833 (at 27°)
Acid value	9.3	0.30	6.73—33.03	0.66
Carvone (%)	55.8	20	—	Nil
Solubility in 80% alcohol	Soluble in 1 vol	Soluble in 1—3 vols	—	Soluble in 1 vol or more

C. EXTRACTIVES[5]

The essential oil of spearmint is obtained by steam distillation. Approximately 0.6% is obtained from the leaves and flowering tops. The odor has been described as having a strong impact with a sharp, fresh, penetrating minty, warm, smooth, pleasantly aromatic, sharp, herbaceous flavor. The oil is a colorless, yellow or greenish yellow liquid with the characteristic odor and taste of spearmint. The aroma improves on aging.

D. VOLATILE OIL[3,12,144-158]

Table 30 gives the characteristics of spearmint oils obtained on an experimental scale at different places in India.

1. Constituents of Volatile Oil[3,13,144,145,148,158]

The characteristic constituent of the volatile oil is *l*-carvone. Oil distilled at Kanpur, India contained carvone 55.8%, terpenes (chiefly *l*-limonene and dipentene) 17.5% alcohol (as dihydrocarveol) 6.7, and esters (as dihydrocarveol acetate) 11.6%. A sample of oil distilled from plants grown in Poona, India contained no carvone; a terpenic glyoxal $C_{10}H_{14}O_2$ was the principal constituent (54%); other constituents of the oil were cineole (31.4%) *l*-limonene, traces of α-pinene, phellandrene, and dihydrocarveol and its ester. The differences in the composition of oils are evidently due to varietal or even specific variations. The essential oil from spearmint cultivated in Italy has been analyzed in detail by Maffei et al.[149]

Spearmint oil produced in the U.S. is derived mainly from Scotch mint (*M. cardiaca*), a species closely related to *M. gentilis* and considered to be a hybrid of *M. arvensis* and *M. spicata*. Russian oil probably obtained from *M. verticillata* Linn var. *strabala* Briq, contains linalool 50 to 60%, cineole .20%, and carvone 5 to 10%.

E. USES[3,13,62,71,145,146]

Spearmint oil is used profusely for flavoring chewing gums, tooth pastes, confectionery, and pharmaceutical preparations. The oil has not attained much commercial importance in India yet even though its impact is felt.

Green leaves of the plant are used for making chutney and for flavoring culinary preparations, vinegar jellies, and iced drinks. The herb is considered stimulant, carminative, and antispasmodic. A soothing tea is brewed from the leaves and an alcoholic beverage (mint julep) is prepared from them as an antidote for certain poisons. A sweetened infusion of the herb is given as a remedy for infantile troubles, vomiting in pregnancy, and hysteria. The leaves are used in fever and bronchitis. Water extract of the leaves when taken orally is documented to relieve hiccups, flatulence, giddiness, and indigestion.

TABLE 31
Comparison of the Physical and Chemical Properties of *Mentha arvensis* (L.)[150,151]

Properties	*M. arvensis* (native)	*M. arvensis* (exotic)
Specific gravity at 20°	0.9329	0.9052
Refractive index at 20°	1.4578	1.4585
Optical rotation	−33.0°	−38.7°
Acid value	0.12	1.00
Saponification value	34.56	51.75
Ester as Menthyl acetate	12.07	18.01
Saponification value after acetylation	203.2	241.75
Free alcohol (as menthol)	46.34%	66.50
Carbonyl (as menthene)	31.63%	11.70
Congealing point	+2.0°C	+15.5°C

III. *MENTHA ARVENSIS* (L.)

Mentha arvensis or Japanese mint is cultivated on a large scale in Jammu and Kashmir and in the Tarai and Haldwani areas in India, covering more than 6000 to 9000 acres. Japanese mint is a downy perennial herb with running roots, a rigid branching stem, 60 to 90 cm high, cultivated at an altitude of 270 to 1500 m. This specie is more robust than *M. arvensis*. It does not breed true from seed. Due to its wide adaptability, it can be cultivated to a large extent. Temperate to tropical climates suit it well. Sunny weather with moderate rainfall is conducive to its luxuriant growth and high menthol content is generally obtained.

A. VOLATILE OIL[150-153]

Steam distillation of the dry leaves gives a yield of nearly 2% essential oil. The physicochemical properties are tabulated in Table 31.

The oil may be profitably used for isolation of menthol if cheaper processes for its isolation other than chilling (which is especially suitable for this purpose) is practicized.

The oil mainly contains menthol 46%, *d*-menthone 15%, carvonmenthene 7.1%, piperitone 2%, methyl acetate 13.2%, limonene 3.7%, and phellandrene 1.2%.

Twenty-eight *Mentha arvensis* seedlings selected randomly and then propagated vegetatively were reported to have produced oils having atypical composition.[155] The pulegone, menthone, and menthol contents of these oils varied markedly. One of the oils contained 2.5% menthol and 84.2% of menthone, a second oil contained 8% menthol and 83.2% pulegone.[155] However, normal values are 80% menthol, 9% menthone, and 1% pulegone. Atypical oils exhibited optical rotations between − 18° and + 26° compared with − 32° and − 43° for the normal oils examined.[155]

The essential oil of *Mentha arvensis* Linn spp. *haplocalyx briquet* var. *piperasceus* Holmes raised at Haldwani, India contains *l*-menthol, *l*-menthone, *l*-pinene, thujone, *l*-limonene, β-phellandrene, cineol, α-thujone, *l*-carvomenthane, pulegone, carvone, *l*-piperitone, caryophyllene, cardinene, a sesquiterpene, and one unidentified oxygenated derivative of sesquiterpene.

B. SPECIFICATIONS[57]

The E.O.A. specifications for *Mentha arvensis* are given in Table 32.

C. USES[1-6,12,62,71]

The dried plant is used as a stomachic, diuretic, and stimulant. It possesses antispasmodic and emmenagogic properties. It is used in jaundice, and is frequently given to stop vomiting.

TABLE 32
E.O.A. Specifications (1966) for the Oil of *Mentha arvensis* L.[57]

	Oil of cornmint / Oil of fieldmint } dementholised
Preparation	By removal of menthol from the whole *mentha arvensis* oil by freezing operations and occasionally rectification
Appearance and odor	Colorless to yellow liquid or characteristic minty odor, the unrectified oil is usually dark yellow to light brown in color
Specific gravity	0.888 to 0.904/25°C
Optical rotation	−20 to −35°
Refractive index	1.4585 to 1.4650
Total alcohol as menthol	40 to 60%
Total esters as menthyl acetate	5 to 20%
Ketone content as menthone	30 to 50%
Solubility	Soluble in 2 to 3 vols of 70% alcohol may become hazy on further dilution
Stability	
Acids	Unstable in the presence of strong acids
Alkali	Unstable due to hydrolysis of esters
Solubility	
Benzyl benzoate	Soluble in all proportions
Diethyl phthalate	Soluble in all proportions
Fixed oils	Soluble in all proportions
Glycerine	Insoluble
Mineral oil	Relatively soluble
Propylene glycol	Slightly soluble

In China, the leaves and stems are made into infusion and used as a carminative and antispasmodic. The plant is also considered as an excellent diaphoretic. An infusion is given in fevers and cephalalgia. The juice of the leaves is applied to the sting or bite of poisonous animals. The leaves pounded with salt are applied externally for wounds. Neither the herb nor the pure menthol can be recommended as anthelmintic. The oil is now coming into use as a flavoring agent also.

IV. *MENTHA LONGIFOLIA* (LINN)[140,156]

A. BOTANY[140]

Mentha longifolia is commonly known as horsemint and is an erect or diffuse herb 30 to 100 cm high with a strong aromatic odor. It is reported to occur in the temperate Himalayas and West Tibet at an altitude of 1200 to 3600 m and in the Kashmir, Garhwal, Kumaon, and Punjab regions of India in large quantities. In India two varieties are distinguished: var. *incana* (wild) and var. *royaleana*.

B. VOLATILE OIL[140,156]

Wild growing plants of *Mentha longifolia* L. showed differences in botanical characters and odor of its essential oil from *Mentha sylvestris* L. with which it is often confused. The physicochemical constants of the oil determined are given in Table 33.

The essential oil was found to contain *dl*-menthol, *l*-menthone, *l*-piperitone, piperitone oxide, aldehyde, *dl*-hydrocarveol, β-caryophyllene, and cadinene. The composition of this oil differs characteristically from the oil of the two known botanical varieties of *M. sylvestris* and this provides specific characters for aid in nomenclature. The taste and flavor of oil were similar to *M. piperita* L. and it was found acceptable for use in the flavoring industry.

TABLE 33
Physicochemical Properties of *Mentha longifolia* and *Mentha sylvertris*[140]

Properties	*M. longifolia*			*M. sylvestris*		
	Sinha & Rajendra Gupta[156]	**Gildemeiss and Hoffman[13]**	**Handa et al.[146]**	**Chopra & Kapoor[157]**	**Bogol et al. in Guenther[13]**	**Kapoor & Chopra[160]**
Yield (w/w) fresh leaves	0.6%	0.57% (dry herb)	1.2% (dry leaf)	0.53%	—	0.63%
Specific gravity 20°/20°	0.9988	0.974	0.9850	0.9969	0.8823	0.18376
Refractive index (20°C)	1.471	1.473	1.471	1.4916	1.4611	1.4742 (30°C)
Optical rotation	– 11°20′	—	—	– 18°4′	– 12°3′	—
Acid value	3.14	4.5	—	1.45	3.5	4.6
Ester value	128.2	127.47	65.8	72.0	147.5	78.67
Ester value after acetylation	218.4	192.7	—	245.0	—	197.27
Carbonyl value as $C_{10}H_{16}$ (hydroxyl amine method)	28.2	38.18	—	4.3	—	3.57

C. USES[1-6,12,156]

Horsemint is considered carminative, antiseptic, and stimulant. The leaves are astringent and used for rheumatic pains. A decoction of the plant is used in fever. The plant is also eaten in the form of chutney.

V. *MENTHA SYLVESTRIS* VAR. *INCANA*[150,151]

The essential oil of this plant raised in the Kumaon region of India and has the following physicochemical properties: specific gravity 0.9895/10°, refractive index 1.495, acid value 8.642, ester value 117.8, ester value after acetylation 243.4, and carbonyl value as $C_{10}H_{14}O$ at 7.2.

The oil has also been examined for its chemical constituents by Baslas and Baslas[151,152] and Sinha and Gupta[156] and the presence of the following constituents has been confirmed: menthol-l-menthane, *l*-piperitone, piperitone oxide, dihydrocarveol, caryophyllene, and cadinene.

VI. *MENTHA PULEGIUM* (L.)[150,151]

The chemical composition of the essential oil obtained (0.9% yield) by Handa et al.[146] is reported as pulegone 73.64%, isopulegone 5.25%, menthone 2.7%, isomenthone 1.86%, piperityl acetate 1%, piperitenol 0.9%, cadinene 0.87%, and an unidentified alcohol 0.84%. The chemical constituents of the different varieties of mint leaves from different sources are tabulated in Table 34.

TABLE 34
The Physicochemical Properties of the Oil of Important Varieties of Mint Leaves[140,146,159-165]

Physicochemical properties	*Mentha longifolia* (hill mint) 1	2	3
Specific gravity	0.9427 (32°C)	0.9027—0.9457	0.9850 (—)
Optical rotation	—	−8°9′—19°42′	—
Refractive index	1.4580	1.445—1.4508	1.471
Acid value	22.09	0.83—13.6	—
Ester value	86.35	56.02—75.61 (11.42 one sample)	65.80
Ester value after acetylation	220.25	167.15—224.79	—
Ester content	30.52	19.80—33.80	—
Total alcohol	73.50	53.70—75.28	—
Free alcohol	44.68	30.05—49.08	—
Ketone content	20.00	15.97—36.94	—
Pheonols content	—	—	—
Solubility in 70% alcohol at 290°C (v/v)	—	1.13—1.21	—

Physicochemical properties	*M. Sylvestris* 1	2	3	*M. piperin* (*Black mitchan*) 1	2
Specific gravity	0.9969 (20°C)	0.8823 (20°C)	0.9837—1.0577 (—)	0.8996—0.9335 (30°C)	0.9565 (30°C)
Optical rotation	−18°4′	−12°3′	+30.69°—7027° (in alcohol)	−11°21′—−16°44′	—
Refractive index	1.4916 (20°C)	1.4611	1.4742—1.4940	1.4380—1.4564 (30°C)	1.4458
Acid value	1.45	3.50	1.47—14.92	0.34—0.91	1.93
Ester value	72.00	147.50	62.01—103.15	26.24—42.27	42.03
Ester value after acetylation	245.00	—	197.27—237.18	194.16—222.92	260.88
Ester content	—	—	—	9.33—14.94	14.86
Total alcohol	—	—	—	61.87—86.37	87.46
Free alcohol	—	—	—	53.57—63.29	77.45
Ketone content	4.30	—	3.69—15.76	18.28—31.48	12.97
Pheonols content	3.50	—	4.30 (one sample)	—	—
Solubility in 70% alcohol at 290°C (v/v)	—	—	1:1.2	—	—

Chapter 15

OREGANO

I. *ORIGANUM VULGARE* L. OR *ORIGANUM* SPP. — FAMILY: LABIATAE

Origanum also called oregano or Mexican sage consists of the dried leaves and flowering tops of a variety of *Origanum vulgare* L. a perennial herbaceous plant belonging to the Labiatae, or mint family. It is indigenous to the sunny, sloping, and hilly areas around the Mediterranean region and western Asia. It is cultivated in many parts of the world including Spain, France, Italy, India, Yugoslavia, Albania, Greece, Turkey, and Mexico. In the past 25 years its use has multiplied rapidly due to popularity of pizza.[3]

A. BOTANY AND CULTIVATION[3,60,70,75]

The *Origanum* plant reaches a height of 1 to 2 ft and has ovate shaped, light green leaves and small purplish flowers borne in clusters on short spikes. Although a perennial, *Origanum* is usually grown as an annual.[1] It grows well in rich and light soils. The plant is quite hardy and can withstand cold weather.[4]

The commercial product (possibly a variety of *Origanum vulgare*) consists of the dried light to brownish green ovate leaves (about 3 cm in length) and the floral parts of the plant. Leaves are broadly ovate, entirely or rarely toothed; flowers are purple or pink in corymbose cymes; nutlets smooth and brown. According to Parry[1] "on both the upper and lower epidermis there are numerous curved, pointed, unicellular nonglandular hairs, some papillose — others smooth, unicellular oil glands with very short unicellular stalks, very small, unicellular capitate hairs with multicellular stalks, and small bicellular, biseriate glandular hairs with unicellular stalks". The stomata are numerous on the lower surface of the leaf.[1] The *Origanum* plant is hardy and can be grown in all warm garden soils. It is usually propagated by seeds, cuttings, layers, and also root division. Dried *Origanum* has a strong aromatic camphoraceous odor and a warm pungent bitter taste. The leaves contain volatile oil, a fixed oil, proteins, and minerals. The plant is harvested when in full bloom. It is commonly sold in bundles often dried. The composition of oregano is given in Table 35.

B. VOLATILE OIL[13,166,167]

The changes in the composition of volatile oil during maturation of oregano have been investigated in detail by Maarse and Van Os[166] and Maarse.[167]

The herb contains a volatile oil of aromatic, spicy, somewhat basil-like odor. According to Guenther,[13] the fresh herb yields 0.07 to 0.20% oil and dried herb 0.15 to 0.40% oil.

The physicochemical properties of the oil vary within the limits as given in Table 36.

The oil contains thymol, carvacrol, free alcohols (about 13%) esters, and 12.5% of a bicyclic sesquiterpene.[13] Several components have been identified by Maarse and Van Os[166] by gas-chromatography and infrared (IR) and mass spectrometry (MS). The major components being sabinene (13.5%), *cis*-β-ocimene (13.5%), caryophyllene (9.2%), *trans,trans*-α-farnesene (2.4%), trans-β-ocimene (3.8%), *p*-cymene (3.1%), linalool (3%), β-bourbonene (4.3%), *trans*-2-hexenal (6.6%), γ-terpinene (1.9%), 3-octanol (1.5%), and other minor components.

II. *ORIGANUM VULGARE* (L.) VAR. *VIRIDE*[3,13,75]

Origanum vulgare L. var *Viride* grows wild in southern Europe and also in the southern part of the U.S.S.R. It contains volatile oil which differs from that of the *Origanum vulgare*

TABLE 35
Composition of Oregano[1-6]
(100 g Edible Portion)

Component	Quantity
Water	7.2 g
Food energy	306 kcal
Protein	11.0 g
Fat	10.3 g
Total carbohydrate	64.4 g
Fiber	15.0 g
Ash	7.2 g
Calcium	1576 mg
Iron	44 mg
Magnesium	270 mg
Phosphorus	200 mg
Potassium	1669 mg
Sodium	15 mg
Zinc	4 mg
Niacin	6 mg
Vitamin A (as β-carotene)	7000 I.U.

TABLE 36
The Physicochemical Properties of the Oil of *Origanum vulgare*[13]

Constants	Oil from U.S.	Oil from Guatemala
Specific gravity at 15°	0.868—0.910	0.952
Optical rotation	−20°0′—−70°0′	+0°25′
Refractive index at 20°	—	1.5084
Ester content	2.0—3.0% (as geranyl acetate)	—
Phenol content	Up to 7.0% (mainly thymol)	68.4%

species by a considerably higher content of thymol. It is possible that the oil is identical with that distilled from *Origanum virens*.

A. VOLATILE OIL[13]

Nearly 1.1% of the volatile oil is obtained from the dried flowering herb. The fresh flowering overground parts of the above-mentioned variety upon distillation yielded 0.35% of oil. The yield of oil from the fresh herb growing wild near the southern U.S.S.R. was nearly 0.4%.[13] Table 37 shows the physicochemical properties of oils of different origin.

The oil of the viride species contains the following compounds: *p*-cymene (17.5%), *d*-pentene (about 10.5%), a terpene, thymol (about 50%), and a sesquiterpene (similar to caryophyllene). In *Origanum vulgare*, carvacrol is the main constituent than thymol.

1. Spanish Oregano Oil

Culzolari et al.[172] examined the chemical composition of an oil of Spanish oregano. They found that it contained α-thujene + α-pinene (1.69%), camphene (0.3%), β-pinene (0.32%), myrcene (2.13%), α-terpinene (2.22%), limonene (0.26%), 1,8-cineole (0.36%), pentanol + unknown (6.5%), γ-terpinene (23.0%), *p*-cymene (12.64%), linalool (0.93%), *cis*-sabinene hydrate (0.58%), caryophyllene (3.32%), borneol (1.42%), β-*bis*-abolene (4.53%), carvone (trace), thymol (18.34%), and carvacrol (4.56%).

This is an unusual oil since oils from *Coridothymus capitatus* normally contain carvacrol as their major constituent. Masada[173] used a combination of GC and MS to characterize the

TABLE 37
Physicochemical Properties of *Origanum vulgare* L. (var *viride*) from Different Countries[13,75]

Constants	Sicily	Italy (Calabria)	Southern U.S.S.R.
Density at 14°	0.9244	0.918	0.9381/20°
Optical rotation	−0°2′	−1°40′	Inactive
Refractive index	1.50029	1.4991	1.5059
Acid number	0.74	—	—
Ester number	2.43	—	—
Ester content	0.85%	—	—
Saponification number	3.17	—	—
Thymol content	50.3%	45.0% (as phenols)	60.0% (as phenols)
Solubility	—	Soluble in 2 vols of 80% alcohol	—

following: α-pinene, camphene, myrcene, limonene, 1,8-cineole, *p*-cymene, 3-octanol, linalool, caryophyllene, geraniol, thymol, and carvacrol in a sample of Spanish oregano oil. Masada[173] also tentatively identified menthone and carvone in the same oil.

Papageorgiou[174] used a combination of capillary GC and mass spectral computer analysis to isolate the constituents of an oil (3.2%) of *Coridothymus capitatus* L. (Syn *Thymus capitatus* L.) which was obtained from plant material of Greek origin. The author found that, in addition to carvacrol, thymol, myrcene, *p*-cymene, and α-terpinene it contained α-pinene, α-thujene, 3-carene, limonene, β-phellandrene, *cis*-ocimene, terpinolene, γ-terpinene, 6-methyl-5-hepten-2-one, 3 octanol, α-*p*-dimethylstyrene, 7-octen-4-ol, *cis*-linalool oxide (furanoid) α-gurjunene, linalool, pulegone, bornyl acetate, caryophyllene, terpinene-4-ol-aromadendrene, isodihydrocarvone, (*cis*-dihydrocarvone) dihydrocarvone, *trans*-pinocarveol, dihydrocarvyl acetate, humulene, neral, camphene, β-pinene, α-terpineol, borneol, carvenone, geraniol, carvone, naphthalene, δ-cadinene, *bis*-abolene cuminaldehyde, nerol, geraniol, *p*-cymen-8-ol, *cis*-6,10-dimethyl-5,9-undecadien-2-one carvacryl acetate, 4-methyl-2-(1,1)-dimethyl-ethyl-phenol, 1-(1,4-dimethyl-3-cyclohexene-1-yl)-ethanoneledol, 4-(1,1-dimethylethyl)-phenol, eugenol, diethylphenol, and 2-methoxy-1,3,5-trimethylbenzene.

Subsequently, Papageorgiou and Argyriadou[175] re-examined the trace constituents (presence in oil less than 0.1%) of the same oil analyzed by them earlier. They reported the presence of 2,5-diethyltetrahydrofuran (no isomer given) hexyl acetate, a butylbenzene isomer, 2-octanone, bornyl chloride, 1-propenyl-2-methyl-1,3-cydohexadiene copaene isogeraniol, 4,8-dimethyl-nona-4,7-dien-2-one, hydrocinnamaldehyde, piperitenone, 2-methyl-4-phenyl-2-butyl acetate, γ-octalactone, 2-(3-pentyl)-phenol, α-calacorene, ledyl acetate, γ-nonalactone, γ-undecalactone, and atisirene (a diterpene hydrocarbon).

The oil of *Coridothymus capitatus* obtained from plants grown in Israel has been examined chemically by Ravid and Putievsky[168,171] and was found to contain α-pinene (2.9%), camphene (0.3%), β-pinene (0.3%), sabinene (0.2%), δ-3-carene (0.1%), myrcene (3.0%), α-terpinene (3.2%), limonene (0.7%), 1,8-cineole (0.6%), γ-terpinene (19.4%), *p*-cymene (6.0%), camphor (trace), linalool (1.0%), terpinen-4-ol (0.8%), caryophyllene (5.0%), borneol (1.2%), thymol (39.3%), and carvacrol (12.7%).

The chemical composition of *C. capitatus* volatile oil was reviewed by Lawrence[169] and the components characterized and compiled by the same author:[169] α-thujene, α-pinene, camphene, β-pinene, myrcene, α-terpinene, limonene, δ-3-carene, *cis*-ocimene, γ-terpinene, *p*-cymene, terpinolene, α-phellandrene, β-phellandrene, α-*p*-dimethyl styrene, α-gurjunene, copaene, β-caryophyllene, aromadendrene, thujone, camphor, 6-methyl-hept-5-en-2-one, 2-undecanone, *p*-menth-3-en-9-al neral, geranial, isogeranial, cuminaldehyde, hydrocinnamaldehyde, hexyl acetate, octenyl acetate, bornyl acetate, terpinene-4-yl-acetate, dihydrocarvyl acetate, β-terpinyl acetate, γ-octalactone, γ-nonalactone, γ-undecalactone, alloaromadendrene, *trans*-α-bergamotene, isocaryophyllene, α-humulene, naphthalene, δ-cadinene, γ-murolene, β-*bis*-abolene, α-calacorene, a butyl benzene, atisirene, pentanol, 3-octanol, 1-octen-3-ol, 1-octen-4-ol, dodecanol, linalool, *cis*-sabinene hydrate, borneol, terpinen-4-ol, *trans*-pinocarveol, α-terpineol, δ-terpineol, nerol, geraniol, ledol, pipertenone, pulegone, menthone, carvone, *cis*-dihydrocarvone, *trans*-dihydrocarvone, carvanone, fenchone, ledyl acetate, carvacryl acetate *cis*-3-hexenyl pentanoate, 1,8-cineole, *cis*-linalool oxide, caryophyllene oxide, bornyl chloride, diethyl phenol, *p*-cymen-8-ol carvacrol, thymol, 2-(3-pentenyl)-phenol thymoquinone, 2,5-diethyltetrahydrofuran eugenol, *cis*-6,10-dimethyl-5,9-undecadien-2-one, 4-methyl-2(1,1-dimethylethyl)-phenol, 4-(1,1-dimethylethyl)-phenol, 1-(1,4-dimethyl-3-cyclohex-1-enyl)-ethanone, 4,8-dimethyl-nona-4,7-dien-2-one, 2-methyl-4-phenyl-2-butyl-acetate, 2-methoxy-1,3,5-trimethylbenzene, 1-properyl-2-methyl-cyclohexa-1,3-diene, *cis*-3-hexenyl-*trans*-4-hexenoate, and 3-methyl-2,6-diisopropyl phenol. He also reported the analysis of four different samples of Spanish *Origanum* oils. The results indicate that in all of the oils examined from Portugal, Greece, Spain, and Crete carvacrol (80%) was the main constituent, the rest being present less than 1% as listed above.[169]

Fleisher et al.[170] reported that thymol, carvacrol, and mixed phenol forms of *C. capitatus* were found growing in Israel. They were found to possess the following phenol contents: thymol types: thymol (50.5 to 67.0%) and carvacrol (3.9 to 7.2%); carvacrol types: thymol (0.6 to 0.7%) and carvacrol (63.0 to 76.1%); and mixed phenol types: thymol (18.0 to 29.6%) and carvacrol (34.5 to 43.7%). Ravid and Putievsky[171] reported the occurrence of two types of *C. capitatus* in Israel.

1. Thymol type: α-pinene (1.6%), camphene (0.2%), β-pinene (0.2%), sabinene (trace), δ-3-carene (trace), myrcene (1.6%), α-phellandrene (0.1%), α-terpinene (1.5%), limonene (0.4%), 1,8-cineole (0.4%), γ-terpinene (9.5%), *p*-cymene (7.4%), terpinolene (0.4%), 3-octanol-(trace), 1-octen-3-ol (0.1%), camphor (0.7%), linalool (0.1%), terpinene-4-ol (0.8%), β-caryophyllene (3.8%), α-terpineol + borneol (0.3%), carvone (trace), bornyl acetate (trace), thymol (52.6%), and carvacrol (18.1%).
2. Carvacrol type: α-pinene (2.1%), camphene (0.1%), β-pinene (0.2%), sabinene (trace), δ-3-carene (trace), myrcene (1.8%), α-phellandrene (0.1%), α-terpinene (1.5%), limonene (0.2%), 1,8-cineole (0.6%), γ-terpinene (12.2%), *p*-cymene (11.3%), terpinolene (0.1%), 3-octanol (trace), 1-octen-3-ol (0.3%), camphor (0.7%), linalool (2.5%), terpinen-4-ol (3.1%), β-caryophyllene (5.9%), α-terpineol + borneol (1.3%), carvone (1.1%), bornyl acetate (0.1%), thymol (9.0%), and carvacrol (39.8%).

III. USES[1-6,16,60,70,71,74,78]

Oregano may be used in all the ways in which sweet marjoram is utilized. But since its flavor is stronger than that of the sweet variety, it should be used carefully according to individual tastes. The herb, either fresh or dried has a wonderful effect upon the flavor of tomato or bean soup when a small spring or a half teaspoon is added to the ingredients. Oregano is also used to flavor sausages; and when boiling or roasting pork or lamb, a sweet pungency is added to the flavor.[1-6]

Oregano is an essential ingredient of chili powder and is used in chili con carne and many other Mexican dishes. It is the spice that made pizza famous. It is equally good in any tomato-type dish, from spaghetti sauce to old-fashioned stewed tomatoes. Oreganum is also used for flavoring soups, meat dishes, pork, fish, egg dishes and salads.[6] Poultry, game, sauces, and vegetables are all more delicious when a touch of oregano has been added. Seafood salads are also flavored with oregano (1 tsp. of ground oregano equals to 0.038 ounce). The leaves are generally crumbled just before using and should not be exposed to high temperatures for long periods of time.

The leaves and tops prior to blooming are used to flavor foods in the same way as sweet marjoram (*Majorana hortensis*). The plant is used in India as a pot herb, and is eaten as a vegetable also.

The oil possesses carminative stomachic, diuretic, diaphoretic and emmengogue properties. It is given as a stimulant and tonic in colic and diarrhea; it is also applied in chronic rheumatism, toothache, and earache. Due to spasmolytic action of the oil, it is used in whooping cough and bronchitis. It is used as an external application in healing lotions for wounds, usually in conjunction with other herbs. It is used for gargle and for bath. The oil is used in the cosmetics and soap industries. It is reported to have stimulating effect on growth of hair.[6]

Chapter 16

PARSLEY

I. *PETROSELINUM SATIVUM* AND *PETROSELINUM HORTENSE* AND *PETROSELINUM CRISPUM* — FAMILY: UMBELLIFERAE

Parsley consists of the dried leaves of *Petroselinum sativum* a biennial herbaceous plant belonging to the Umbelliferae, or carrot family (also known as the parsley family). The parsley plant is believed to be native to Sardinia. It has been grown in England since the middle of the 16th century and has become naturalized in the U.S. since then and also in India. Parsley is extensively cultivated throughout Europe and North America.[3,4,6,74,122,176] In the U.K. and U.S. the curly-leaved varieties are grown almost exclusively, but in other countries the plain-leaved varieties are more usual and the curly-leaved almost unknown. It is also widely cultivated in the Mediterranean region.

A. BOTANY[3,50,60]

Although parsley is a biennial it is grown as an annual, since its foliage is the principal harvest. Propagation is by seed planted in beds in the early spring with germination requiring about 6 weeks. It is planted 1 ft apart and at 6-in. intervals. About 16 lb of seeds will plant about an acre. It requires moist clay and loam soil.

The plant gives a profusion of bright green leaves, segmented and curly. The flowers produced in the second year, are small, white and borne in compound terminal umbels. Under cultivation, parsley is usually grown as an annual and will produce several crops of leaves. Two to five cuttings of leaves are possible for each planting before flowering.

Parsley thrives in normal good soil in the temperate and subtropical regions. It is a smooth biennial herb with many branched stems growing as high as 65 cm at times and bearing divided leaves which in some varieties are greatly curled and crisped.

The fresh leaves are dried rapidly under controlled conditions of heat and air to avoid loss of the essential oil that gives the herb its characteristic flavor and ensures good green color.

B. DEHYDRATION[3,60,74]

Dried parsley usually has a bright green color. Its volatile oil is readily lost, particularly during drying. Microscopically, parsley does not show either hairs or crystals. The upper epidermal cells are polygonal but those on the lower epidermis are wary. The epidermal cells tend to be striated. The characteristic green color of the leaves can be retained well if the leaves are dried rapidly. About 12 lb of de-stemmed fresh parsley are required to produce 1 lb of the dehydrated product. The U.S. production is about 3 million lb of dried leaves per year.

The approximate composition of dried parsley is shown in Table 38.

The leaves, stems, and fruits contain a glucoside apiin which on hydrolysis yields aprigenin, glucose, and a sugar; apiose, a second glucoside, consisting of luteolin, glucose, and apiose has also been reported. The fruits of plants from some parts of northern India yielded 2.2% of apiin, but the second glucoside was not detected.

C. COMMERCIAL SPECIFICATIONS IN THE U.S.[5]

The product consists of the leaf portion of freshly harvested parsley, that has been washed, trimmed, diced, dried, and size classified. When reconstituted, its flavor shall be fresh, typical of fragrant parsley, and its color shall be uniformly bright green. A maximum of 20% of the flakes may pass through a U.S. standard No. 20 sieve. The dehydrated product

TABLE 38
Composition of Dried Parsley, 100 g Edible Portion[5,18]

Component	Quantity
Water	8.0—9.0 g
Food energy	276 kcal
Protein	22.4 g
Fat	4.0—4.4 g
Total carbohydrate	51.7 g
Fiber	10—10.3 g
Ash	12—12.5 g
Calcium	1468 mg
Iron	98 mg
Magnesium	250 mg
Phosphorus	350 mg
Potassium	3805 mg
Sodium	425—450 mg
Zinc	5 mg
Ascorbic acid	120 mg
Riboflavin	1 mg
Niacin	8 mg
Vitamin A (as β-carotene)	23340 I.U.

shall be practically free from yellow or discolored leaves. Its moisture content shall not exceed 4.5%. To preserve its freshness, the dehydrated product should be packed in heat sealed, polyethylene lined, corrugated fiber cases and stored under dry, cool storage conditions at 70°F or below. Alternately, the product may be packed in either hermetically sealed metal containers or in 5-gal metal containers with friction top lids.

D. EXTRACTIVES[5]

The plant yields two different types of essential oils: one from the whole plant and one from the seed. The yield of oil from the freshly harvested plant is not more than 0.06% whereas that from the seed is as high as 3.5%. The flavor is much harsher and more herbaceous than the seed oil which is somewhat viscous, yellow-amber in color, and has a warm, bitter, aromatic taste.

The oleoresin of parsley has a flavor characteristic of the entire plant and is noticeable even in small percentages of it. It is a deep green, semiviscous liquid, containing 12 to 15 ml of volatile oil per 100 g. One hundred fifty grams (1/3 lb) are equivalent to 45.45 kg (100 lb) of freshly harvested parsley.[5]

E. VOLATILE OIL[3,6,13,177]

All parts of the plant contain an essential oil which is responsible for the characteristic aroma and flavor of parsley. The oil is recovered by steam distillation and is used mainly for flavoring of food products. The oil obtained from the flowering tops is of the finest quality, truly representing the odor of the leaves, but the yield is too small (less than 0.05%) for commercial products. Commercial parsley oil is distilled either from the aerial parts of the herb bearing immature fruits (herb oil, yield about 0.25%) or from the mature fruits (fruit oil, yield 7%). There is considerable difference in the physicochemical characteristics of the herb and seed oil. The physicochemical properties of the different types of parsley herb oil are given in Table 39.

The oil is found to contain more than 42 components.[176] The main constituent of the volatile oil is the phenol ether apiole, known also as parsley camphor, which crystallizes in

TABLE 39
The Physicochemical Properties of the Different Types of Parsley Herb Oil[1,3,13]

Characteristics	French	American	Hungarian
Specific gravity 15°	0.911—1.002	0.909—1.046	0.948—0.987
Optical rotation	−6°10′—+6°0′	−7°40′—2°13′	−6°18′—2°3′
Refractive index at 20°	1.5029—1.5159	1.5080—1.5179	1.5053—1.5250
Acid number	1.4	—	0.1—0.2
Ester number	0.9—8.9	—	2.0—6.1
Ester number after acetylation	26.1—50.4	—	11.5—22.2
Saponification number	—	0.0—6.1	7.4—7.7
Solubility	95% alcohol soluble with opalescence	0.5 vol of 90% alcohol	3.5 vol of 90% alcohol

the form of long needles and is colorless. It is a 2,5-dimethoxy, 3,4-methylene dioxy-1-allyl benzene. In the French type of parsley, the 2,5-dimethoxy is replaced by 5-methoxy and the compound myristicin is the constituent.

The constituents of parsley essential oil are myristicin, α-pinene, small amounts of aldehydes, ketones and phenols, a luteolin monomethyl ester, chlorogenic acid, caffeic and ferulic acid; furocoumarins like bergapten, xanthotoxim, imperatoxim, glycoside, aspirin, luteolin *m*-apio-sylglucoside, luteolin-7-diglucoside, moderate concentrations of α-tocopherol and *p*-mentha-1,3,8-triene; others include pinenene, β-phellandrene, 1,3,8-*p*-menthatriene, terpinolene, and 1-methyl-4-isopropyl benzene.[176]

Sunderman[178] showed the presence of UDP apiose in the UDP sugar fraction from parsley by a method comprising extraction with ethyl alcohol and chromatography with Dowex 1-XZ (formate). The UDP sugars reported were D-glucose 77% (of UDP sugars); D-galactose 21%; xylose 4.5%; *l*-arabinose 2%; *l*-rhamnose 2%; and fructose 0.35%. The presence of an alkaloid has also been reported. If the oil contains more than a certain quantity of myristicin it becomes semi-solid at ordinary temperatures. However, crystals are deposited on exposure to cold.

1. Adulteration[1,3,13]

The herb oil is usually adulterated with the seed oil which is of low quality. This can be noted by the changes in the rotation from dextro for leaf oil exclusively to levo for seed oil.

F. USES[1-6,12,52,74,176]

Both the dried and the fresh herb have innumerable uses. The fresh herb is the one used for garnishing and seasoning and is also eaten. The leaves are also employed to make tea.

The acid sweet pungency of parsley flavor is known worldwide and its beauty as a garnish, is almost taken for granted. The dried leaves of parsley (known as parsley flakes) are used in flavoring soups, meat, and fish dishes, vegetable dishes, and salads. It is also used an an ingredient in meat and poultry seasoning. An unusually delicious use of the herb is made in preparing biscuits.

Parsley juice is rich in Vitamin C, carotene, iodine, and in most of the valuable organic salts. It is useful as a general tonic and is known to act specifically as a tonic for kidneys. In minute doses apiole is of service as a curative of epileptic fits. Leaves applied to the breasts several times a day is said to suppress secretion of milk effectively. Bruised leaves are used as a poultice for sore eyes. Parsley tea is reported to have carminative and diuretic properties and aids in digestion.[5,6,71]

The herb is reported to posses diuretic, carminative, emmenagogue, and antipyretic properties, and has long been in use for uterine troubles.[3,6] Bruised leaves are applied to bites and stings of insects, and the mericarps are used to get rid of lice and skin parasites. The juice of the fresh leaves is used as an insecticide. Parsley causes skin reactions in some people, and this is attributed to the presence of a furocoumarin and bergapten.

Chapter 17

ROSEMARY

I. *ROSMARINUS OFFICINALIS* (L.) — FAMILY: LABIATAE

Rosemary is often found near the coastal areas where it thrives under conditions of fog and salt spray. However, it also grows inland. Most of the rosemary today is grown commercially in France, Spain, Portugal, Yugoslavia, Germany, Italy, Morocco, Romania, Russia, North Africa and some is produced in the U.S.[3] The plant also needs winter protection and thrives in cool climates. The flavor of the leaf is strongly associated with the type of climate it grows in. It is also cultivated in Indian gardens in cool places for its pleasantly fragrant leaves. However, production of this leafy spice is increasing.

A. BOTANY AND CULTIVATION[1,3,13,16,51,57,60,70-72,74-78]

Rosemary is a shrubby, sweet-scented perennial which is often treated as an annual in the colder climates. This aromatic herb has tall, erect, branching stems which reaches up to more than 6 to 7 ft in height. It is an evergreen bush with pale blue flowers and spiky leathery leaves. The leaves are narrow, about 2 to 5 cm long and resemble curved pine needles. The obtuse, slender, linear leaves, are green on top and grayish green and cottony looking on the underside. They have narrow caves with inrolled margins. The small bluish pink or almost lavender flowers blossom in clusters at the end of the tall branches. The spicy, blended aroma and flavor of the leaves, flowers, and stems makes this shrub of southern Europe one of the best-loved of all the herbs. By the end of the second year after seedling, the rosemary plant is a dense shrub 2 ft in diameter.

The chief structural characteristics of the leaves are, they are narrow, hairs thin-walled, hairy beneath, branching, sometimes with globular heads on the branches, also jointed capitate, and the bladder glands eight-celled. The cells of the upper epiderm are larger and more often beaded than those of the lower epiderm.[69] If the leaves are to be used for culinary purposes, they must be dried as soon as possible after being harvested to avoid the loss of the volatile oil. When dried carefully on trays in a well ventilated, dark, sheltered place, the leaves retain their green color and their clean, fresh, bitter-sweet flavor. The dried herb is available whole or ground.[1,57,64] The leaves become rolled in appearance dark-green to brownish-green in color and have a tealike fragrance. When crushed, rosemary has an agreeable fragrant, aromatic, eucalyptus-like aroma with a cooling, camphoraceous scent. The taste is somewhat peppery, spicy, herbaceous, warming, and astringent with a tint of bitterness.[5]

B. SPECIFICATIONS IN THE U.S.[5]

Rosemary leaves shall be dried, clean, whole leaves of *Rosmarinus officinalis* L. The shiny, dark green to brownish green colored, rolled margined leaves shall have the shape and appearance of pine needles. Rosemary has a tealike fragrance; when crushed, the leaves have a slightly camphoraceous odor and a somewhat bitter and slight camphoraceous taste. The product shall contain no more than 7.0% total ash, 0.5% acid insoluble ash, 7.0% moisture, and not less than 1.1 ml of volatile oil per 100 g of 95% of the ground product, and must pass through a U.S. standard No. 35 sieve.[5]

The approximate composition of dried rosemary leaves is shown in Table 40. It is rich in calcium and potassium as well as carbohydrates.

C. VOLATILE OIL[13,55,72,75,179]

The amount of volatile oil available from the leaves varies within the general range of the family up to 2% or more. The leaves of fresh rosemary plants constitute 70% of the

TABLE 40
Composition of Dried Rosemary Leaves (100 g Edible Portion)[5]

Component	Quantity
Water	9.3 g
Food energy	331 kcal
Protein	4.5—4.9 g
Fat	14—15.2 g
Total carbohydrate	64.1 g
Fiber	17.7 g
Ash	6.0—7.0 g
Calcium	1280 mg
Iron	29 mg
Magnesium	220 mg
Phosphorus	70 mg
Potassium	955 mg
Sodium	50 mg
Zinc	3 mg
Ascorbic acid	61 mg
Niacin	1 mg
Vitamin A (as β-carotene)	3128 I.U.

structure, the stalks 30%. The oil yield from the leaves is 0.1% and from the stalks it is 0.05%. Therefore, recovery from the stalks is uneconomical. Steam distillation is superior to water distillation (0.08% recovery based on fresh plant). Hexane extraction of the dried leaves gives a yield of 1.6% concrete oil out of which 24.1% is absolute oil. The total quantity of oil decreases gradually from shade to sun drying. Shade drying of the plants for 2 days prior to oil recovery is considered optimum, since water loss is 30%, while oil loss is only 1.64% (of oil present). Nearly 200 kg of leaves are necessary to obtain 1 kg of essence (at 0.5% yield).

1. Composition of Volatile Oil[5,130,179,180]

Chemically, the constituents of oil are 16 to 20% borneol, 27 to 30% cineole, 10% camphor, 2 to 7% bornyl acetate, and small percentages of α-pinene, camphene, terpinol, and verbenone.

Borneol is responsible for the pungent, camphoraceous odor and burning taste, camphor contributes to a cooling, penetrating, minty note. Cineole gives it a fresh, cooling eucalyptus-like aroma, and α-pinene is responsible for its warm, piney notes; bornyl acetate accents the fresh, sweet piney notes while adding its herbaceous characteristics.

The oleoresin is a greenish-brown, semisolid 10 to 15 ml of volatile oil per 100 g. Generally, 2.5 kg of oleoresin are equivalent to 45.5 kg of crushed and dried rosemary leaves.[5]

Oil of rosemary (obtained by distillation from flowering tops or leafy twigs of *R. officinalis*) is made up of nearly 60% monoterpenes like pinene, 13% borneol and camphor, and 3% of esters of borneol, other minor constituents present are camphene and cineole, free alcohols (10%) including borneol and linalool, and alcohols such as eucalyptol (about 12%). The amount of esters calculated as bornyl acetate and of total borneol, respectively varies with geographical source of oil of rosemary.[13,60,75]

The physicochemical properties of the oil produce in India (at Doddabetta) near Ooty in South India and other oils from various places are given in Table 41.[179,181]

The percentage of esters and alcohols according to Virmani et al.[181] of different origins are given in Table 42.

TABLE 41
Physicochemical Properties of the Oil of *Rosmarinus officinalis* of Different Origins[13,181]

Characteristics	Indian oil	Spanish oil	French oil	Yugoslavian oil
Specific gravity	0.8966 at 20°	0.893—0.910 at 25°C	0.900—0.920 at 15°C	0.894—0.913 at 15°C (mostly above 0.9020)
Optical rotation	+9°28′	−0°58′—+11°30′ at 25°C	Up to +13°10′ and in isolated cases slightly levorotatory	+0°43′—5°53′ and only in very rare cases is higher
Refractive index	—	1.4682—1.4712 at 20°C	1.467—1.472 at 20°C	1.466—1.468 at 20°C
Ester content as bornyl acetate	2.79%	1.0—3.9 (in an exceptional case as high as 5.4)	1.0—4.9%	1.8—7.0%
Total alcohol content as borneol	11.42%	8.7—15.6	8.0—11.3	8.4—14.3
Solubility	Positive in an equal vol of 90% alcohol positive in 10 vols 80% alcohol; solution clear	Positive in 4.5 to 5 vols of 80% alcohol many genuine oils turbid in 10 vols of 80% alcohol	Usually positive 1 to 8 vols of alcohol	Usually soluble in 1 to 8 vols of 80% alcohol

TABLE 42
Esters and Alcohols Present in Rosemary Oil of Different Origin[181]

Origin	Esters (as bornyl acetate) (%)	Alcohols (as borneol) (%)
French	1—4.9	8—9
Spanish	1—3.5	10—14
English	4.9	13.7

The following monoterpene hydrocarbons are determined by silicic acid chromatography followed by gas-liquid chromatography:[13] α-pinene, α-thujene, camphene, β-pinene, sabinene, δ-3-carene, α-phellandrene, α-terpinene, myrcene, *d*-limonene, β-phellandrene, γ-terpinene, *p*-cymene, terpinocene, and ocimene. The triterpene alcohols according to Brieskorn and Doembling[182] are *epi*-α-amyrin, β-amyrin, betalin, and β-sitosterols. Carnosic acid, a derivative of fermginol is shown to be a major terpenic constituent of *Rosmarinus officinalis*.[13,66,69] Carnosic acid is converted oxidatively into rosmarcine carnosol and other compounds of similar structure. The occurrence of 5-hydroxy-4′-7-domethoxy flavone in *Rosmarinus officinalis* leaves has been confirmed by Brieskorn and Doembling.[182] Its structure was confirmed by UV and NMR spectra.

Tucker and Maciarello[180] have grouped rosemary cultivars into six chemotypes based upon the composition of their essential oils. The major components (α-pinene, camphene, 1,8-cineole, camphor, bornyl acetate, and borneol) vary widely in percentage composition. The total varieties listed are 23 in number. Some of the values such as light blue to the very oily type and Benender blue and Tuscan blue have exceptionally low percentages of camphor. On the other hand, Majorca pink, Gorizia, and Joycede Buggio have very low α-pinene content. Both Gorizia and Joycede Buggio have low 1,8-cineole. An exceptionally high content of 1,8-cineole is found in the Asp cultivar.

The exhaustive review by Boelens[183] on essential oil of rosemary with nearly 93 references is recommended for the reader who would like to know more details about rosemary essential oil. The quantitative analyses of Spanish, French, and different origins of rosemary have been reported in this review in detail. Also, Boelens[183] has dealt in detail with the biogenesis and chemical formation, genetic interpretation, and reconstitution of rosemary oil. Lawrence in a later review has tabulated most of the identified compounds till 1985. The components identified by various workers are[184-189] α-thujene (2.7%), α-pinene and toluene (21.8%), camphene (5.6%), β-pinene (3.1%), sabinene and mesityl oxide (2.0%), 3-carene and myrcene (2.8%), α-phellandrene (0.7%), 3-hexanone and limonene (3.7%), 1,8-cineole (19.0%), *trans*-ocimene and γ-terpinene (0.2%), *p*-cymene (1.0%), terpinolene (0.1%), methyl heptenone (0.2%), 3-octanol (0.1%), α-*p*-dimethyl styrene and *trans*-sabinene hydrate (0.2%), longifolene and α-fenchyl acetate (1.0%), 3,5,5-trimethylhexan-l-ol (0.7%), camphor and linalool (19.8%), iso(iso)pulegol (0.1%), *p*-menth-3-en-l-ol, α-fenchyl alcohol and bornyl acetate (2.1%), terpinen-4-ol (0.4%), caryophyllene (0.4%), *cis*-β-terpineol (0.5%), isopinocarveol (0.4%), δ-terpineol, lavandulol, isoborneol, *trans*-β-terpineol and α-humulene (0.9%), α-terpineol, borneol, α-amorphene, and γ-muurolene (7.0%), verbenone (0.9%), myrtenol (0.2%), nopol, calamenene, *trans*-carveol and *p*-cymen-8-ol (0.1%), caryophyllene oxide (0.1%), elemol, benzyl alcohol, and γ-eudesmol (0.5%). In addition, trace amounts of ethanol, α-terpinene, β-phellandrene, 1-octen-3-ol, copaene, δ-cadinene, γ-cadinene, *cis*-carveol and calacorene were also isolated and identified. A laboratory distilled oil was analyzed and was found to contain α-thujene and α-pinene (12.5%), camphene (4.0%), β-pinene (1.3%), δ-3-carene and myrcene (1.3%), α-terpinene (0.4%), limonene (3.0%), 1,8-cineole (47.0%), γ-terpinene (0.4%), *p*-cymene (1.8%), copaene (0.6%), camphor and linalool (10.7%), bornyl acetate (0.9%), terpinen-4-ol-(1.3%), caryophyllene (4.9%), δ-terpineol (0.9%), borneol, α-humulene, and α-terpineol (4.0%), γ-muurolene (3.1%), β-*bis*-abolene and carvone (0.4%), and aromatic curcumene (0.2%). In addition, trace amounts of sabinene, terpinolene, α-p, dimethyl styrene, α-fenchyl alcohol, verbenone, ledene, δ-cadinene, γ-cadinene, α-selinene, cubenene, *p*-cymen-8-ol, *trans*-anethol, calmenene, calacorene, caryophyllene oxide, methyl eugenol, humulene epoxide I, humulene epoxide II, α-corocalene, carvacrol, thymol, and cadalene.

2. Adulteration of Rosemary Oil[13]

Oil of rosemary is frequently adulterated with oil of terpentine, but lately other oils are also employed for this purpose, for example, certain fractions of camphor oils, oil of pine steam distilled, eucalyptus oil terpenes, and fractions resulting from the manufacture of terpineol. Strict control of these analyses of oil is suggested.

D. USES[1-6,13,16,54,60-64,68-70,124,126,127]

Oil of rosemary is used as a perfume mainly in rubifacient liniments. The bulk of the oil is used for scenting of soaps and other technical products. Oils having a harsher note such as Spanish and Moroccan oils are used in soaps and denaturing of alcohol whereas oil with its finer, more delicate odor such as Dalmatian oil is preferred for the flavoring of all kinds of food products.[127] Superior grades of the oil are employed for blending in perfumes.

The fresh tops of rosemary may be used as garnishes for fruited summer drinks.[3] It is one of the most fragrant and pleasing of spices and may be used with any vegetables particularly peas, green beans, asparagus, broccoli, cauliflower, potatoes, eggplant, summer squash, zucchini, spinach, and turnip. The freshly chopped leaves are used judiciously in sauces, stems used in soups and roasts have a pleasing effect upon the ultimate flavor. Rosemary blended with chopped parsley and melted butter and spread over a caper before roasting has an excellent taste. The fresh leaves are used in fruit salads, jams, sweet sauces, cream soups, and biscuits.[16,68]

The dried herb has just as exciting a flavor as the fresh leaves, and the unusual tastiness of the Italian and Polish sausages owes to the amount of rosemary which has been blended in with the other seasoning. It is also used in herb vinegars.[69]

The oil also finds use in medicinal preparations and as an ingredient in rubefacient liniments. It is mildly irritant and has been used as a carminative. Internally, the oil may be taken as a stimulant in doses of a few drops: a 5% tincture is used as a circulatory and cardiac stimulant. The oil is useful for relieving headache. It is diaphoretic and is employed with hot water in chills and colds. An emulsion prepared from the oil is used as a gargle for sore throat. The oil exhibits antibacterial and protistocidal activity.[6]

All parts of the plant are astringent and serve as a nervine tonic and an excellent stomachic. An infusion of the plant with borax is employed as a hair wash and said to prevent premature baldness. The plant has been found useful in atonic dyspepsia.

Dried leaves are smoked for the relief of asthma. A decoction of leaves is said to be employed as an abortifacient. Pressed juice of the leaves possesses a strong antibacterial action on *Staphylococcus aureus, Escherichia coli*, and *Bacillus subtilis*. Flowering tops and leaves have a camphor-like odor and taste and are considered carminative, diaphoretic, diuretic, aperient, emmenagogue, stimulant, and stomachic. They are used for the relief of rheumatism, paralysis, and incipient catarrhs.

In recent years, rosemary has been one of the most inexpensive of all the spices and herbs. The bulk of rosemary oil, however, is used for scenting soaps and technical products, and for denaturing of alcohol. The oil is useful in room sprays and inhalants.

Chapter 18

SAGE

I. *SALVIA OFFICINALIS* (LINN) — FAMILY: LABIATAE

Sage has been most widely cultivated as a spice plant for many centuries because of its aromatic odor and unique bitter and pungent taste. The gardens of most European villages, both ancient and modern are filled with its spicy aroma.[3]

Common or garden sage is a native of the Mediterranean region, with approximately 550 species in its genus. *Salvia* are native to the southwestern U.S. and Mexico. It grows profusely over the hillsides of the shores of southern Europe. It is cultivated in various parts of continental Europe including Spain, Italy, Yugoslavia, Greece, Albania, Argentina, Germany, France, Malta, Turkey, England, and also Canada and the U.S.[1-6]

Yugoslavia is a great producer of leaves, herbs, and essential oils of this plant. The islands along its coast produce these products from sage. Among them Austria, Dalmatia, Herzegovina, and Montengaro (Izrua Gora) are the most important exporters. The greatest quantity of dried leaves as herbs and essential oil of the sage is exported to the U.S. under the trade name 'Jugoslav sage' (*Salvia jugoslavica*).[190]

Yugoslavia mainly exploits only wild sage. It is not cultivated as it is so abundant that it can never be completely exploited. Recently, plantations have sprung around the Kashmir valley in India and the quality of leaves and oil obtained from it is comparable with many other sources.

Most of the sage imported into the U.S. in recent times comes from Albania, Italy, and the Dalmatian coast of Yugoslavia. Dalmatian sage is highly esteemed.

A. BOTANY[1,3,17]

Garden sage (*Salvia officinalis* L.) is 2-ft tall, much branched, stiff upright, long-lived semishrub. It has slender, grayish silvery-green, long-petioled shiny oblong leaves of 1 to 2 in. in length, which are with a pebbley texture above. It has pale blue, streaked flowers in spikes at the extremities of the branches. The plant flowers in early summer.[17] All sages have characteristically four sided square stems, flowers in whorls with a two-lipped calyx and two-lipped corolla. It is essential that stock plants with heavy thick leaves should be selected and planted so that the plant displays robust growth. Seeds as a rule do not germinate well and poor strands are usually the result. Sage seems to be perfectly adapted to the climate and soil of the growing sections in Georgia and Tennessee in the U.S.

Sage should be planted from 15 to 24 in. apart in the row and 2.5 to 3.5 ft between rows. The distance depends on the fertility of the soil. The more fertile the soil, the larger the plant, hence the need for more growing space. Several thousand plants are required to fill 1 acre.

The first year it will be a small crop. The second and succeeding years nearly 1200 to 2000 lb/acre of dried sage leaves can be expected from two cuttings. Sage is not subject to any pathogens other than nematodes. Care should be taken, therefore, to avoid the use of land in which these organisms are known to exist. The size and vigor of the plant are much increased by mulching.

B. HARVESTING[12,13]

Garden sage can be harvested during the fall of the first year. Two or three annual harvestings can be made thereafter.

The highest grade product is obtained by harvesting only the leaves. A much cheaper method, involving cutting the stems 3 to 5 in. long or that part of the stems on which the

TABLE 43
Composition of Ground Sage, 100 g Edible Portion[3,5,14,15]

Component	Quantity
Water	6.0—8.0 g
Food energy	315—415 kcal
Protein	10.2—10.6 g
Fat	12.7—14.7 g
Total carbohydrate	50—60.7 g
Fiber	16.1—18.1 g
Ash	7.7—8.0 g
Calcium	1652 mg
Iron	28 mg
Magnesium	428 mg
Phosphorus	91 mg
Potassium	1070 mg
Sodium	11 mg
Zinc	5 mg
Ascorbic acid	32 mg
Niacin	6 mg
Vitamin A (as β-carotene)	5900 I.U.

leaves are growing provides an excellent product. Still, cheaper methods have been practiced, but they give a much lower quality product. One of these methods is cutting the crop with a moving machine, with the cutter bar set at the proper height; the sage is then gathered into piles with a hay rake. The first harvesting is usually done when the plants begin to bloom. A second and third harvesting can be made between this period and frost time, depending upon the growing season.

Collection of the sage is preferable in the afternoon, since at that period of the day the flowers and leaves contain the maximum amount of essential oil. The flowers contain more essential oil (average 25%) than the leaves, therefore, one must gather both flowers and leaves at the same time.[190]

C. DEHYDRATION

The herb gardener usually cures his sage by cutting it towards the end of the summer and then hanging the reapings in bunches, in a dry, airy place. A lower quality product will be obtained from air-dried sage (spreading the crop thinly in the shade) since drying should be continuous from the time the foliage is harvested and will give a darker-colored product which has less value in the market.

Dehydration by modern methods is the most satisfactory manner of curing sage for the market and gives the highest quality product. If a dehydrator is employed, the temperature should be started at 100°F and carried on for 4 hr. This usually gives the best quality product which will vary slightly with the moisture content of the foliage. The moisture should be reduced to less than 13%. It is said that cultivated sage bears more sand than wild-growing material; it must, therefore, be thoroughly washed before drying.[3] Immediately, following dehydration, the product should be shipped to market without any delay as the loss of volatile oil is continuous. Shipping sage in cleaned fertilizer or feed bags is satisfactory and much cheaper than in boxes or crates.[3]

D. CHEMICAL COMPOSITION OF SAGE[3,5,14-16,19]

The chemical composition of sage is shown in Table 43.

E. SPECIFICATIONS IN THE U.S.[5]

Sage shall be the dried leaves of *Salvia officinalis* L. with green to gray-green, oblong, lanceolate leaves, covered with fine short hairs that possess a strong, fragrant, and aromatic odor, free of any camphoraceous note. It shall contain a maximum of 10% by weight, of stems, excluding petioles. The sage shall contain not more than 10.0% total ash, 1.0% acid insoluble ash, and 8.0% moisture. It shall contain not less than 1.0 ml of volatile oil per 100 g and not less than 95% of the ground sage shall pass through a U.S. standard No. 20 sieve. For rubbed sage, 95% shall be retained on a U.S. standard No. 40 sieve and at least 95% shall pass through a U.S. standard No. 20 sieve.[5]

F. ADULTERATION OF LEAF[3,13]

Adulterants occasionally found in *Salvia officinalis* include the leaves of *S. lavandulaefolia* and *Salvia triloba* (Greek sage). The leaves of *S. lavandulaefolia* are oblong lanceolate and smaller than *S. officinalis* and tend to be whorled on the stems. The apex is either acute or rounded, the base subcordate or rounded, the margin nearly netire, very slightly crenutate, and the surface much smoother than that of *S. officinalis*. Similarly, Greek sage leaves are broader, shorter, thicker more wooly, with very short petioles, and with a less pronounced crenulate margin; the odor of Greek sage is characteristically aromatic. It contains up to 2.37% of volatile oil and is stronger than common sage, but the flavor is much inferior than Dalmatian. The most common adulterants in *S. officinalis* are the sage stems. These may be detected microscopically by the presence of numerous reticulate vessels and crystal-bearing cells.

G. EXTRACTIVES[5]

The essential oil, to which the plant owes most of its flavor and character, is produced by steam distillation of the freshly harvested leaves. The yield is about 2.5%, though it could be much less depending on the climatic conditions of the growing and harvesting seasons and the country of origin. Its color is pale yellow to almost colorless. Its odor is strongly aromatic, initially cooling then warming, spicy, herbaceous, camphoraceous, and very persistent with a burning flavor, eucalyptus-like, with a lingering, bitter, sweet note. The oil from Greek sage has a harsher, more eucalyptic aroma while English sage has an odor with more camphoraceous characteristics. Spanish sage has an odor that is somewhere between the odor of the English and Greek sages.

The oleoresin of sage is usually prepared from the Dalmatian type of sage, *Salvia officinalis* L. The oleoresin has a warm, spicy taste, a very important attribute for sausage seasonings. The oleoresin is brownish-green and has a very heavy liquid consistency. It contains a minimum volatile oil content of 25 to 30 ml per 100 g. Nearly 3.4 kg of oleoresin is equivalent to 45.5 kg of freshly ground sage in aroma and flavor characteristics.[5] The oleoresin may be used in conjunction with the corresponding essential oil in the flavoring of all kinds of food products.[5]

H. ESSENTIAL OIL OF SAGE

The various parts of the sage plant have been analyzed for essential oil by Tucakov.[190] They were taken from the same plants, at the same time, and analyzed by the same method. Stems proved to be the poorest in essential oil, especially the underground part average value of 0.15%.

In the flowering plants (stems, leaves, and flowers) an average of 0.60% essential oil is reported. The leaves of the flowering plant contain an average of 0.64%, the flowers about 1.10%; from leaves analyzed from plants before flowering values of 1.14 to 2.28% up to 2.47% (average 1.80%) are reported.[3]

The highest quantity of essential oil is reported in the leaves of plants which had not yet flowered. On an average a value of 1.90% is reported. Hence, distilleries are mostly interested in the leaves of sterile plants and those picked from fertile ones prior to flowering. Leaves gathered from the sterile plant contain about 10% more essential oil than those of the fertile ones. Also the increase in the amount of oil is greatest in the morning. This increase is more rapid in the leaves of plants without flowers.[3,190]

Influence of the temperature and method of drying on the percentage of essential oil in the leaves of sage is fairly well documented.

The least loss of essential oils is reported by open air driers and those of IR rays at a temperature not exceeding 30°C. Drying in direct sun resulted in as high as 24% loss of essential oil.[13] In shade-drying the losses are between 2 and 1%. A temperature of 25 to 30°C was the best, both regarding the prevention of undue loss of oil and in giving the product a good appearance with low moisture content.

1. Physicochemical Properties of Volatile Oil[3,13]

The important physicochemical properties of volatile oil from different origins are given in Table 44.

2. Composition of Volatile Oil

Chemically, the essential oil contains 40 to 60% thujone, 15% cineol, up to 16% borneol, up to 4% bornyl ester, α-pinene, salvene, and *d*-camphor. In Dalmatian sage the following components are reported: salvane ($C_{10}H_{18}$), *d*-α-pinene, cineole, *d*-β-thujone, *l*-α-thujone, borneol, and *d*-camphor.

Guenther[13] reported the following components: terpenes (15.0%), ketones (31.5%), cineole (15.0%), free alcohols (11.2%), esters (2.3%), and sesquiterpenes (20.3%).

The volatile oil from different parts of sage are given in Table 45.[69]

Ivanic and Sarin[192] have compared the physicochemical properties of various sage oils distilled from *S. officinalis* obtained from Sicevska klisura, Montenegro, and Dalmatia. Subsequently, Ivanic et al.[193] compared the chemical composition of the oils from the above regions of Yugoslavia with Dalmatian sage.

Thujone is the main constituent of Dalmatian sage oil and is isolated from the fraction at 198 to 203°C. It was originally called salviol; *d*-β and *l*-α thujones were identified as semicarbazones with m.p. 174 and 186°C, respectively. Thujone -$C_6H_6(C_3H_7)$-(C)(CH_3) is responsible for the characteristic flavor of sage oil.

Ursolic acid was isolated by Brieskorn and WesCamp from sage.[194] The leaves of *S. officinalis* on extraction with boiling 96% EtOH, condensation of the extract, washing off the precipitate with chloroform and hot water, and on recrystallization from 96% EtOH yielded 2.1% ursolic acid. A more soluble form of ursolic acid named ursolic acid II has been isolated from the chloroform wash. Oleanolic acid was also isolated from this solution after the removal of ursolic acid II by alkali extraction.[192] Brieskorn et al.[192,195] obtained two modifications of ursolic acid (*cis* and *trans*). Oleanolic acid was also obtained in two forms, the α-acid, m.p. 309°C and the β-acid m.p. 297°C. A neutral crystal fraction m.p. 228 − 30°C was also recognized. The terpenoids in *S. officinalis* (Dalmatian) were determined and separated by gas chromatography by Brieskorn et al.[195] It contained 3-methyl-3-methylene-5-heptane and farnesene.

Picrosalvin an aromatic diterpene, *O*-diphenollactone is the bitter principal of the sage reported by Brieskorn et al.;[192,195] nearly 120 mg of picrosalvin can be obtained from 30 kg of sage. The optical activity of the essential oil of sage is due to camphor.[191]

Burgar et al.[196] used a combination of gas chromatography and ^{13}C-NMR to characterize the presence of the main components in Yugoslavian sage oil. The compounds identified were as follows: α-thujone (28.0%), β-thujone (14.5%), camphor (18.0%), 1,8-cineole (14.0%), camphene (3.7%), α-pinene (3.5%), α-humulene (2.9%), β-pinene (2.8%), bornyl

TABLE 44
Physicochemical Properties of Volatile Oils of Sage of Different Origin[3,13,191]

Property	Yugoslavian	American	Turkey	Poland	India	Appenine Zone
Yield of oil	0.7—1.4%	—	1.0%	—	1.1% (yellow)	0.5—0.6% (yellow)
Specific gravity (20°)	0.915—0.927	0.922—0.926	0.890	0.6—0.9218 at 15°	0.9268	0.818
Optical rotation (20°)	+28°56′—11°38′	+4°28′—+4°56′	+24.4°	+2.74°	0.2°	+1.6°
Refractive indices	1.4571—1.4758	1.4637—1.4699	—	—	—	—
Ester content (borneol acetate)	1.6—4.9%	3.3—6.0%	—	—	—	—
Total alcohol content (borneol)	6.9—16%	13%	—	—	—	—
Ketone (thujone)	22.0—61.2%	35.41—46.7%	—	—	—	—
Saponification number	—	—	—	—	—	11.06
Ester value	—	—	—	—	30.1	—
Ester value after acetylation	—	—	—	—	30	—
Ester number	—	—	17.2	6.48	—	—
Acid value	—	—	—	—	1.1	—
Solubility in alcohol	9.5 vol of 80% 7—10 vol of 70%	1 vol of 80% 5.5 vol of 70%	—	—	—	3 vols of ethyl alcohol
Acetylation number	—	—	—	33.2	—	—
Acid number	—	—	1.4	0.37	—	—

TABLE 45
Contents of the Volatile Oil of Flowering and Nonflowering Plants of *Salvia officinalis*[13]

Contents	Flower plant (%)	Nonflowering plant (%)
Cineole	32—35	13.0—20
Sesquiterpenes	About 30	About 20
Camphor and thujone	5.0—10.0	20.0 to 32
Borneol	9.0—14.0	7.5 to 12.0
Terpenes	—	About 15.0
Esters	2.0	2.2 to 3.7

acetate (2.7%), borneol (2.2%), linalool (1.8%), β-caryophyllene (1.7%), limonene (0.7%), farnesene (0.6%), carvone (0.6%), linalyl acetate (0.6%), salvene (0.5%), myrcene (0.3%), tricyclene (0.3%), and α-thujene (0.1%).

Marlier et al.[200] used gas chromatography and mass spectrometry to examine the chemical composition of an oil derived from *Salvia officinalis* grown in Peru. The components identified by them are α-pinene, camphene, β-pinene, 3-carene, myrcene, α-terpinene, limonene, 1,8-cineole, *trans*-3-hexenal, γ-terpinene, *p*-cymene, terpinolene, hexanol, 3-hexenol, α-thujone, β-thujone, 1-octen-3-ol, camphor, isopinocamphene, linalool, bornyl acetate, terpinen-4-ol, α-humulene, *trans*-pinocarveol, α-terpineol, and borneol.

Kustrak et al.[199] reported that a standard quality Yugoslavian sage oil should possess the following composition: α-pinene (1.3%), camphene (2.75%), β-pinene (4.80%), sabinene (2.12%), myrcene (1.39%), α-terpinene (0.21%), limonene (1.68%), 1,8-cineole (10.93%), γ-terpinene (0.62%), *p*-cymene (0.57%), α-thujone (24.57%), β-thujone (13.98%), camphor (12.35%), isobornyl acetate (3.39%), bornyl acetate (0.93%), and borneol (4.42%).

Grzunov et al.[204] used thin-layer chromatography to tentatively characterize the presence of thujyl β-D-glucoside and thymol β-D-glucoside in an acetone extract of sage leaves. Alcohols of these same sugar esters are found in sage oil.

Vernin and Metzger[202] compared the chemical composition between Albanian and Dalmatian sage oils. The contents of the various components in percentage for Albanian and Dalmatian oil respectively are α-thujene (0.17, 0.12), α-pinene (3.40, 3.97), camphene (4.90, 3.19), β-pinene (1.10, 0.97), sabinene (0.10, 0.10), myrcene, (0.55, 0.72), α-terpinene (0.08, 0.14), limonene (1.46, 1.25), *cis*-ocimene (0.04, 0.02), γ-terpinene (0.11, 0.18), *trans*-ocimine (0.02, 0.01), terpinolene (0.11, 0.12), *p*-cymene (0.8, 0.12), 1,8-cineole (8.2, 12.0), α-thujone (23.4, 37.15), β-thujone (3.45, 14.20), camphor (22.45, 12.30), linalool (0.51, 0.40), linalyl acetate (traces), bornyl acetate (3.5, 0.86), β-caryophyllene + terpinene-4-ol (6.5, 2.2), α-humulene (6.9, 3.87), borneol + α-terpineol + α-terpinyl acetate (5.30, 1.87), *p*-cymen-8-ol (0.13, 0.05), and humulene oxide (0.80, 0.48).

The authors[202] also identified trace amounts of the following compounds (in either Albanian or Dalmatian sage oils or both): hexane methylcyclopentane, *cis*-2-methyl-3-methylene hept-5-ene, *trans*-2-methyl-3-methylene hept-5-ene, 1,8-cineole, sabinyl acetate, isoborneol, δ-selinene, α-muurolene', δ-cadinene, 1-octen-3-ol (Z)-3-hexenyl isovalerate, (E)-sabinene hydrate, (E)-3-hexenyl isovalerate, isocaryophyllene, aromadendrene, alloaromadendrene, carveol, geraniol, caryophyllene oxide, eugenol, thymol, carvacrol, and humulenol or caryophyllenol.

Rhyu[197] examined the chemical composition of a number of lab-distilled *Salvia triloba* oils by gas chromatography. The compounds identified with the percentage range of each compound are α-pinene (4.1 to 6.0%), camphene (2.2 to 4.0%), β-pinene (2.3 to 5.3%), sabinene (trace), α-thujone (1.0 to 2.8%), camphor (1.5 to 2.6%), α-terpinene (0.3%), limonene (1.5 to 2.3%), myrcene (1.5 to 3.0%), γ-terpinene (0.2%), *p*-cymene (1.2%), terpinolene (0.1%), 1,8-cineole (40.7 to 54.1%), linalool (7.0 to 16.8%), β-caryophyllene (5.0 to 7.6%), and α-terpineol (1.4 to 2.5%).

Tucker et al.[198,203] have indicated the correct taxonomic identification of *S. triloba* (Greek sage) is *Salvia fruticosa* Mill. Putievsky et al.[205] compared the chemical composition of *Salvia fruticosa* Mill. (syn. *S. triloba*). They found that the major components varied slightly from organ to organ in the following way (as shown in parentheses): α-pinene (18.6 to 37.3%), camphene (2.1 to 3.1%), β-pinene (5.0 to 6.6%), myrcene (0.1 to 3.4%), β-caryophyllene (7.6 to 11.3%), 1,8-cineole (30.8 to 44.0%), camphor (3.3 to 6.8%), borneol (1.0 to 1.3%), and α-terpinyl acetate (2.0 to 3.4%).

Bayrak and Akgul[206] used retention time data to characterize the presence of the major constituents in *S. fruticosa* oil (produced from plant material harvested in Mugla, Turkey). The constituents identified were as follows: α-pinene (3.2%), camphene (0.7%), β-pinene (4.3%), myrcene (3.1%), α-terpinene (0.4%), limonene (trace), 1,8-cineole (55.5%), γ-terpinene (1.7%), terpinolene (0.9%), α-thujone (0.2%), β-thujone (2.0%), camphor (8.4%), linalool (1.6%), linalyl acetate (0.05%), bornyl acetate (1.7%), borneol (4.6%), terpinene-4-ol-(1.1%), β-caryophyllene (5.2%), α-humulene (0.3%), and α-terpinyl acetate (1.1%).

Harvala and co-workers[207] examined the chemical composition of a lab-distilled oil of *S. triloba* collected from the island of Cephallonia, Greece. The summary of their results are α-pinene (2.56%), camphene (2.68%), β-pinene (2.75%), myrcene (1.95%), terpinene-4-ol (1.18%), β-caryophyllene (4.13%), α-humulene (0.79%), borneol + α-terpineol (7.29%), γ-terpinene (0.20%), limonene + 1,8-cineole (38.51%), γ-terpinene (0.51%), *p*-cymene (0.75%), terpinolene (0.15%), 1-octen-3-yl-acetate (0.02%), α-thujone + 1-octen-3-ol (2.34%), dehydro-*p*-cymene (trace), β-thujone (4.12%), camphor (15.15%), linalool (1.27%), linalyl acetate (0.48%), bornyl acetate (0.96%), α-terpinyl acetate + naphthalene (2.19%), carvone + bisabolene (0.43%), neryl acetate (trace), δ-cadinene + gernalyl acetate (0.36%), myrtenol (0.10%), nerol (0.07%), calamenene (0.16%), geraniol (0.07%), calarene (0.14%), caryophyllene oxide (0.91%), thymol (0.11%), and carvacrol (0.33%). In addition, the authors also found that the oil contained two sesquiterpene alcohols that could not be characterized.

Bayrak and Akgul[206] used retention indices to identify the major constituents in a lab-distilled oil obtained from constituents in *S. officinalis* cultivated in Izmir (Turkey). The compounds identified were as follows: α-pinene (3.5%), camphene (4.7%), β-pinene (1.8%), myrcene (0.8%), α-terpinene (0.1%), limonene (1.4%), 1,8-cineole (5.0%), γ-terpinene (0.4%), *p*-cymene (1.1%), α-thujone (20.6%), β-thujone (15.1%), camphor (22.9%), linalool (1.1%), linalyl acetate (1.0%), bornyl acetate (2.6%), borneol (7.9%), β-caryophyllene (4.1%), α-humulene (2.1%), α-terpinyl acetate (0.9%).

Recently, Lawrence[208,209] examined the chemical composition of sage oils produced from *S. officinalis* cultivated in an experimental garden in the U.S. A summary of the range of composition of the major volatile constituents identified spectroscopically is as follows: *cis*-2-methyl-3-methylene hepta-5-ene (tr. to 0.1%), α-thujene (tr. to 0.2%), α-pinene (1.68 to 5.37%), camphene (1.67 to 5.74%), β-pinene (10.5 to 14.48%), sabinene (tr. to 0.79%), myrcene (0.27 to 1.09%), α-terpinene (tr. to 0.32%), limonene (0.55 to 2.46%), 1,8-cineole + *cis*-ocimene (2.26 to 18.90%), *trans*-ocimene (tr. to 1.67%), γ-terpinene (tr. to 0.97%), *p*-cymene (0.10 to 1.05%), terpinolene (tr. to 0.34%), α-thujone (13.94 to 44.05%), β-thujone (2.47 to 9.90%), camphor + α-jurjuene (1.94 to 21.14%), linalool (0.16 to 0.54%), bornyl acetate (0.19 to 1.41%), caryophyllene + terpinen-4-ol (1.50 to 10.05%), α-humulene (1.53 to 10.65%), α-terpineol (tr. to 0.28%), sabinol (1.30 to 7.66%), δ-cadinene (tr. to 0.1%), caryophyllene oxide (0.16 to 1.27%), virdiflorol (1.35 to 9.91%), and manool (0.11 to 5.19%).

As can be seen from all these results, the chemical composition of *S. officinalis* oil can vary considerably depending upon the source.

TABLE 46
E.O.A. Standard (1966) for Oil of Sage (Dalmatian — *Salvia officinalis* L.)[57]

Properties	Specifications
Preparation	Obtained by steam distillation of the partially dried leaves
Physical and chemical contents	
Color and appearance	A yellowish or greenish-yellow liquid having a warm camphor and thujone-like odor and flavor
Specific gravity 25/25°	0.903—0.925
Optical rotation	+2°—+29°
Refractive index 20°C	1.4570—1.4690
Saponification value	6—20
Ester value after acetylation	25—60
Ketone content	Not less than 50%, calculated as thujone
Solubility in alcohol	1 vol of 80% alcohol
Descriptive characteristics	
Solubility	
Benzyl benzoate	Soluble in all proportions
Fixed oils	Soluble in all proportions in most fixed oils
Glycerine	Practically insoluble
Mineral oil	Soluble, although the solutions are frequently opalescent
Propylene glycol	Slightly soluble
Stability	Relatively stable to dilute weak alkali and weak organic acids
Containers	Should be shipped in glass or tin-lined containers; good quality galvanized containers are suitable when long storage is not contemplated
Storage	Store preferably in tight, full containers, in a cool place protected from light

Jimenez et al.[210] studied the hypoglycemic activity of infusions and suspensions of Spanish sage (*Salvia lavanduifolia*). The authors found that in alloxandiabetic rabbits the administration of 0.250 mg/kg of an infusion daily resulted in a 33% reduction in blood glucose levels.

This same year, Crespo et al.[211] examined the chemical composition of an oil of *S. lavandulifolia* spp. oxyodon harvested in the province of Granada. A summary of the changes that occur during the vegetative growing season is as follows: α-pinene (4.1 to 10.6%), camphene (4.9 to 11.9%), β-pinene + 3-carene (4.3 to 8.9%), myrcene (2.8 to 7.0%), limonene (3.6 to 5.5%), 1,8-cineole (10.3 to 31.6%), *p*-cymene (0.6 to 1.8%), terpinolene (tr. to 0.7%), α-thujone + β-thujone (0.7 to 1.5%), camphor (16.0 to 38.9%), terpinene-4-ol (tr. to 2.8%), borneol (tr. to 23.8%), α-terpineol (tr. to 1.5%), thujyl alcohol (tr. to 2.2%), and α-ionone (tr. to 1.2%).

3. Adulteration of Oil[3,13]

Thujone is the common adulterant of Dalmatian sage oil which can be isolated from the much lower priced American cedar leaf oil. Since thujone is also the main constituent of Dalmatian sage oil, such additions are difficult to prove by mere routine analysis. There are specifications and standards laid down earlier for these which are given in Table 46.

I. USES[1-6,12,55,60,70,74,212,218]

Sage is a very popular herb for use in culinary preparations in the West. It is used in pork sausage and baked meatloaf. Sage is an important herb in every kitchen for flavoring meat and fish dishes and in making poultry stuffings. It is used in poultry dressing, sausage, liver sausage, and hamburger seasoning.

Sage has been extensively employed in the food industry as a standard spice in making stuffing for fowl, meats, and sausage. Dried and powdered leaves are mixed with cooked vegetables and sprinkled on cheese dishes, cooked meats, and other similar preparations. Fresh sage leaves are used in salads and sandwiches. The young leaves are pickled and are also used for making tea.

Sage oil finds use in perfumes as a deodorant, in insecticidal preparations, for the treatment of thrush and gingivitis, and as carminative. The oil is used as a convulsant, and it resembles wormwood oil in action but is less active.[1-6,12]

Sage and sage oil exhibit antioxidant properties principally due to polyhydric phenol.

Sage is used as a mild tonic, astringent, and carminative. An infusion of the leaves is used as a gargle in the treatment of sore throats; hot infusion is said to be diaphoretic. Extracts of sage leaves are also reported to be antipyretic. A strong infusion of the herb is reported to be antipyretic and known to dry up breast milk for weaning children. Sage has been prescribed to cure female disorders since ancient times and estrogenic substances are said to have been extracted from the dried leafy tops. Leaves rubbed on teeth are known to function as a good dentrifice. The herb has been used in toothpastes and mouthwashes, gargles, poultices, tooth powders, hair tonics, and hair dressings.[3,6]

II. CLARY SAGE (*SALVIA SCLAREA* L.)

A. BOTANY[3]

Clary sage is a perennial native to Mediterranean countries, that grows wild in southern France, North Africa, parts of Italy, and also on the Crimean Peninsula in the U.S.S.R. The green part of the plant, and especially the flowering tops, contain an essential oil of delightful odor, somewhat wine-like.[3]

The different types of clary sages are the French, the Russian, the Italian, the Hungarian and the English clary sage.

The oil can be extracted by steam distillation of the flowering tops and leaves of the plant or by extraction with volatile solvents. Table 47 gives an idea of the properties of the oil of different origins.[13]

B. CHEMICAL COMPOSITION

l-Linalyl acetate, linalool, *l*-nerolidol acetate of nerolidol, a sesquiterpene castonyl compound, cedrene, and scalreol are reported to be present in clary sage oil.

C. SPECIFICATIONS

The early E.O.A. specifications are described in Table 48.

D. USES

The oil blends especially well with lavender and bergamot, and also with jasmine. It has also been employed in flavors, especially with wine and other liquors.

III. SPANISH SAGE[5] (*SALVIA LAVANDULAEFOLIA* VAR. *S. HISPANORUM*)

The components of the oils are described earlier. Spanish sage oil is sometimes adulterated with oils of camphor, eucalyptus, steam distilled pine oil, or with fractions of these oils.

A. SPECIFICATION

The standards and specifications laid down earlier for the Spanish sage oil are given in Table 49.

TABLE 47
Physicochemical Properties of Clary Sage of Different Origin[13]

Property	French	Russian	Italian	Hungarian	English
Specific gravity at 15/15°	0.900—0.910	0.900—0.903	0.8930—0.8990	0.907	0.904—0.915
Optical rotation	11°22′—32°38′	−11°22′—15°20′	−15°12′—29°3′	14—14°40′	−16°0′—53°0′
Refractive index	1.4613—1.4700	1.4573—1.4612	1.4675—1.4710	1.4617	1.469—1.492
Ester content (linalyl acetate)	54.0—70.2%	65.7—71.2%	36.58—45.34%	71.2%	34—53%
Evaporated residue	5.3—13.1%	1.6—6.4%	—	4.4%	—
Solubility in alcohol	0.5 vol of 90% alcohol	1 vol of 80% alcohol	0.6—1.2 vol of 80%	1 vol of 80%	1—10 vol of 90%
Acid number	—	—	0.61—1.20	—	0.5—1.8
Ester number	—	—	104.5—129.60	—	—
Saponification number	—	—	105.18—130.8	—	98—164
Total alcohol	—	—	60.2—68.81%	—	—

TABLE 48
E.O.A. Standard (1966) for Oil of Clary Sage (*Salvia sclarea* L.)[57]

Properties	Specifications
Preparation	By steam distillation of flowering tops and leaves of clary sage plant
Physical and chemical constants	
Color, odor, and appearance	Pale yellow to yellow liquid, with a herbaceous odor and a winey bouquet
Specific gravity 25/25°	0.886—0.929 (correction factor 0.006/°C)
Optical rotation	−6— −20°
Refractive index (20°C)	1.4580—1.4730
Acid value	Not more than 2.5
Ester content	48—75% calculated as lanalyl acetate
Solubility in alcohol	Soluble up to 3 vols of 90% alcohol becoming opalascent on dilution
Descriptive characteristics	
Stability	
Alkali	Unstable to alkalies due to the saponification of ester
Acid	Due to high content of rinalyl acetate in clary sage oil, this oil is not stable in acids, since strong acids transform the ester into terpenica product
Solubility	
Benzyl benzoate	Soluble
Fixed oil	Soluble in all proportions in most fixed oils
Glycerine	Insoluble
Mineral oil	Soluble up to 3 vols, becomes opalascent on further dilution
Propylene glycol	Insoluble
Containers	Ship preferably in glass, aluminum, tin-lined or good quality galvanized iron container
›rage	Store preferably in tight full containers in a cool place protected from light

B. USES

Spanish sage oil is occasionally found as an adulterant in lavender oil. When exported it serves primarily for the scenting of soups, usually in combination with lavender, spike lavender, and rosemary.

IV. *SALVIA LEUCANTHA*[3,75,213]

A. BOTANY

S. leucantha is a shrub with 1.5 to 2.5 ft high branches with deciduous leaves short-petioled 2 to 6 in. long, lanceolate-linear acute, crenate, tugose, pubiscent above, tomentose beneath, floaral bracts ovate acuminate, deciduous racemes 6 to 10 in. long whorls many fold calyx, funnel-shaped with about equal lobes, violet to lavender lanate, corolla, white 5/8 to 3/4 in. long swollen at throat, without a hairy ring inside, with upper lip much exceeding layer. The plant flowers during August to October.

In Kumaon hill in India five species namely *S. meorcrofliana, S. glutinosa, S. lanata, S. nubicola*, and *S. leucantha* are widely distributed.[213]

B. CHEMICAL COMPOSITION

The essential oil from the flowering tops of the plant is extracted by steam distillation (yield 0.09%).

C. PHYSICOCHEMICAL PROPERTIES OF THE OIL[3]

The various properties are $d_4^{20} = 0.8650$; $n_D^{20} = 1.476$; $\alpha_D^{20} = +0.5$; acid value 3.013; ester value 51.98; ester value after acetylation 88.2, and carbonyl value 24.2 (as $C_{10}H_{16}O$).

TABLE 49
E.O.A. Standard (1966) for Oil of Sage (*Spanish — Salvia lavandulaefolia*) described by Guenther[57]

Properties	Specifications
Physical and chemical constants	
Color, odor, and appearance	Colorless to slightly yellow oil having a camphoraceous odor with a cincole top note
Specific gravity 25/25°	0.909—0.932 (temp. correction factor 0.00046/°C)
Optical rotation	−3°—+24°
Refractive index at 20°C	1.4680—1.4730
Saponification value	14—57
Saponification value after acetylation	56—98
Solubility in alcohol	Soluble in 2 vols of 80% alcohol which may become opalascent upon dilution
Descriptive characteristics	
Solubility	
Benzyl benzoate	Soluble in all proportions
Fixed oils	Soluble in all proportions in most fixed oils
Glycerine	Soluble in all proportions
Mineral oil	Soluble in all proportions but with opalascence after 2 vols
Propylene glycol	Soluble in all proportions but with opalascence
Stability	
Alkali	Moderately stable in dilute alkali
Acids	Unstable in the presence of strong mineral acids
Containers	Ship preferably in glass or tin-lined containers; good quality galvanized containers are suitable when long storage is not contemplated
Storage	Store in tight, full containers, in a cool place protected from light

The essential oil mainly contained 1,8-cineole, *p*-cymene, β-thujone, linalool, geranyl acetate, citral b, citronellal, citronellol, geranil, cadrene, and aromadendrene.[3]

Around 1976 the essential oil called 'Essence Stenophylla' made its appearance. About 1980, a brief description of its ingredients was published.[140,214,215] *Salvia stenophylla* is a well-known folk medicine in Southern Africa and is used profusely mixed with tobacco also.[215]

The plant is hardy, herbaceous, and erect and grows wild in southern Africa. It grows up to a height of 60 cm and is covered with very short, stiff, glandular hairs and a small number of spherical oil cells. Leaves are linear to elongated and lanceolate. They have pubescence of very short glandular hairs, mostly on the veins of the underside of the leaf. The flowers are of pale-blue or pink in color. It flowers through summer. The plant is found at an altitude of 400 to 2300 ft.

Brunke and Hammerschmidt[214] have analyzed the oil and the physical constants are $d_4^{20} = 0.9075$, $n_D^{20} = 1.4860$, $\alpha_D^{20} = +3.8$. The smell is fresh, herby, and camphoric. The constituents of the oil have been identified by gas chromatography-mass spectrometry.

V. OTHER SPECIES

Among the *Salvia* species growing wild in Israel, *S. fruticosa* (*S. triloba*) and *S. dominica* are of great importance as potential sources for commercial herbs and essential oils. *S. fruticosa* is a native shrub in the Mediterranean basin, 1 to 1.5 m high, which flowers from March until June. Tea prepared from fresh or dried leaves is commonly used as a remedy for stomach pains, coughs, and colds.[168,219-221] The main components in the oil distilled from leaves are 1,8-cineole, camphor, α-pinene, β-pinene, borneol, and β-caryophyllene. The amounts of α- and β-thujone, the major components in *S. officinalis* are were below 0.1%. The proportion of the monoterpene hydrocarbons in the oil was 28.4% and the other constituents are α-thujene (tr.), α-pinene (6.3%), camphene (6.9%), β-pinene (5.9%), sabinene (tr.), myrcene (4.1%), α-terpinene (tr.), limonene (2.9%), 1,8-cineole (46.9%), *cis*-β-ocimene (tr.), γ-terpinene (1.2%), *trans*-β-ocimene (tr.), *p*-cymene (0.8%), terpinolene (0.3%), α-thujone (tr.), β-thujone (tr.), camphor (13.0%), bornyl acetate (1.1%), β-caryophyllene (3.0%), α-humulene (1.0%), and borneol + terpinyl acetate (5.4%).

The essential oil of a spontaneous hybrid of *S. fruticosa* and *S. officinalis* was found to have a unique composition;[222] 1,8-cineole (20%), camphor (19%), α- and β-thujone (16%) were the main components. The oil obtained by steam distillation from freshly cut leaves of *S. fruticosa* was a colorless liquid with a harsh, spike rosemary, camphor-like odor; *S. fruticosa* oil from Turkey was found to be rich in 1,8-cineole (42.%), camphor (9%), β-caryophyllene (8%), and β-pinene (5.5%); it grows wild in Sicily, Greece, Turkey, Lebanon, and Algeria.[223] The major components of the essential oil of sage Greek origin were 1,8-cineole (64%), camphor (8%), β-pinene (3%), and α- and β-thujone (5%). The essential oil from commercial leaves contain 27.6 to 54.1% 1,8-cineole.[197] This compound was also found in high concentration in some North American *Salvia* species.[224]

Salvia dominica[214] is a strong-smelling plant growing wild in chalky hills; the inflorescences and the leaves are rich in essential oil. The odor and composition of the oil resemble the oil distilled from the inflorescences of *S. sclarea*. The monoterpene hydrocarbons are found in low concentration in the oil (6.5%). The oil is rich in monoterpene alcohols and their acetates. The major components are linalool (26%), linalyl acetate (28.4%), and α-terpineol (12.4%).

The characteristic aroma of the oil of *S. sclarea* is due to microcomponents other than linalyl acetate (60%) and linalool (19%).[225,226] The same holds true for the oil of *S. dominica*. Although *S. dominica* oil is not known in commerce, it has a great potential as a substitute or additive to *S. sclearea* commercial oil.

The genus *Micromeria* is common in the Mediterranean basin. *M. fruticosa* grows wild on chalky rocks and flowers from February to June. Tea prepared from the fresh leaves has a minty aroma and is used in folk medicine for indigestion, coughs, and colds and for lowering high blood pressure.[219,229] The main components in the oil distilled from fresh plants are pulegone (45.9%), menthol (10.1%), and limonene (8.0%). The proportion of the monoterpene hydrocarbons was 14.4%. Pulegone, the minty component of *M. fruticosa* is the industrial starting material of menthone and menthol.[10]

Chapter 19

SAVORY

I. *SATUREIA HORTENSIS* L. (SUMMER SAVORY) — FAMILY: LABIATAE

Summer savory, *Satureia hortensis* the savory of commerce, is a small slender herbaceous annual of the mint family indigenous to Southern Europe and the Mediterranean region. There are a number of species of savory, most of them coming from the Mediterranean area. The chief ones are the so-called winter savory and summer savory (more common). Summer savory is used in cooking although in Spain another species, *Satureia thymbra*, more close to thyme in flavor is used as a kitchen herb. The summer savory is mostly cultivated in France and the Mediterranean region, whereas the winter savory (*Satureia montana* L.) is grown in southern Europe. Winter savory has gained popularity in Canada, Germany, and the U.S., the U.S. imports about 80 to 85 tonnes annually.[3,6]

A. BOTANY[60,74]

Summer savory has slender, hairy, erect, branching stems that grow to 15 in. in height. It has a well-developed tap root, an erect stalk, and is pubescent through the longest hairs occurring on the margins of the lower half of the leaf. The leaf is entire, narrow-pointed at the tip, and tapering at the base to the scarcely indent petiole. The leaves are soft dark green 1/4 to 1/2 in. long. Under a lens, the leaves appear finely pilted owing to the depressions beneath the bladder glands. Numerous lavender flowers, up to 5 mm broad and 5 mm long as short pedicles, occur in groups of three in the axis in the leaves. The deeply five-cleft calyx is half the length of the whole flower.

Seed of the summer savory, which is an annual, is sown in the spring, harvesting is carried out about 3 months after planting, just before the plants bloom. The plant will be attractive, slender, fragrant, with tiny tubular, pale-pink and lilac colored flowers.

B. HARVESTING[60]

The leaves are harvested before flowering 75 to 120 days after sowing and are dried. The dried leaves, brownish green in color fold upwards at the midrib. It has a fragrant, aromatic odor and a warm aromatic slightly sharp taste. The taste is somewhat sharp and camphoraceous. It is marketed both as whole leaf, dried and ground form.

C. SPECIFICATIONS IN THE U.S.[5]

Savory shall be the whole or ground dried leaves and flowering tops of *Satureia hortensis* L. The pale, brownish-green leaves have a fragrant, aromatic odor, and a warm, slightly sharp taste. It contains no more than 10% total ash, 2.0% acid insoluble ash, and 10.0% moisture. It contains not less than 25 ml of volatile oil per 100 g and 95% of the ground product will pass through a U.S. standard No. 40 sieve.

D. COMPOSITION[5]

Table 50 gives the approximate composition of dried savory leaf. It is a rich source of calcium and vitamin A. The fat content is considerably low compared to other leafy spices, i.e., rosemary. Analysis of the green herb (leaves and flowering tops) collected at the end of the blooming period showed moisture 72%, protein 4.2%, fat 1.65%, sugar 2.5%, fiber 8.6%, and ash 2.11%.

TABLE 50
Composition of Sweet Summer Savory (100 g, Edible Portion[5])

Component	Quantity
Water	9.0 g
Food energy	270—360 kcal
Protein	6.7—7.7 g
Fat	5.2—5.9 g
Total carbohydrate	54.6—68.7 g
Pentosans	12 g
Fiber	15.3 g
Ash	8.7—9.6 g
Calcium	2132 mg
Iron	38 mg
Magnesium	377 mg
Phosphorus	140 mg
Potassium	1051 mg
Sodium	24 mg
Zinc	4 mg
Niacin	4 mg
Vitamin A (as β-carotene)	5130 I.U.
Vitamin C	12 mg

TABLE 51
The Physicochemical Properties of Oil of *Satueria hortensis*[13,227]

Property	Guenther[13]	Manjunatha et al.[227]
Color and appearance	—	Golden yellow liquid
Specific gravity 15°	0.941	0.986/22°
Optical rotation	2°48′	−1°12′
Refractive index	1.5044/20°	1.495/25°
Saponification value	—	5.3
Saponification value after acetylation	—	147.5
Phenol content (carvacrol)	55.0%	38.0%
Solubility	3—3.5 vols of 70% alcohol	1.5 vols and more of 80% alcohol

E. VOLATILE OIL OF SAVORY[1,3,56,227]

The volatile oil content of savory is highest during full flowering but its fluctuation from bud formation to the end of flowering is small. Generally, the yield of oil is between 0.1 to 0.15% of the air-dried herbage. The oil is golden yellow in color with characteristic thyme-like or oregano-like odor.

The physicochemical properties as reported by Guenther[13] and Manjunatha et al.[227] are given in Table 51.

F. CONSTITUENTS OF VOLATILE OIL[1,3,56,74,75,228,229]

The chief odorous constituent of the volatile oil is carvacrol (30 to 45% depending upon variety). The carvacrol content of essential oil remains constant during flowering.[13] Other organic compounds in the volatile oil include *p*-cymene (173 to 175°C) and probably pinene and dipentane. Cymol and a terpene are the other constituents of the oil. The presence of ursolic acid (0.1%) and etheral oil (0.4%) has been shown by Brieskorn et al.[228]

The known antioxidant action of savory is found to be due in part to the labiatic acid which it contains. The presence of flavonol (which may impart some antioxidant activity) has also been reported.[229]

Leone and Angelescue[230] identified limonene (14%), *p*-cymene (27%), and thymol (28 to 65%) in an Italian oil of savory. Subsequently, Igolen and Sontag[231] found that a French oil of savory contained *p*-cymene (33.7%), linalool, nerol, geraniol, α-terpineol, terpinen-4-ol- and borneol (25%), camphor + carvone (0.3%), dihydrocuminyl alcohol (trace), thymol + carvacrol (30.4%), and some unidentified acids including acetic acid (0.43%).

Later Karawya et al.[232] found that savory oil obtained from plants grown in Egypt contained α-pinene, sabinene, β-pinene, camphene, β-phellandrene, myrcene, *p*-cymene, limonene, 1,8-cineole, camphor, α-terpineol, borneol, linalool, bornyl acetate, and geraniol. Pellecuer[233] reported that French savory oil contained γ-terpinene (6.85%), *p*-cymene (26.26%), linalool (2.85%), α-terpineol (4.28%), terpinene-4-ol (3.42%), and thymol (49.46%). Subsequently, Yugoslavian oil of savory was reported to contain α-pinene (1.75%), β-pinene (2.20%), limonene (2.25%), *p*-cymene (15.04%), 1,8-cineole (5.40%), α-terpineol (5.51%), borneol (4.74%), thymol (6.46%), and carvacrol (49.76%).[235,236]

Pellecuer and Garnero[235] compared the chemical composition of savory oil obtained from several areas in France. The authors identified α-pinene (0 to 2.6%), camphene (0 to 1.0%), β-pinene (0 to 0.8%), myrcene (0 to 2.0%), α-terpinene (0 to 2.4%), limonene + 1,8-cineole (0 to 3.8%), γ-terpinene (0 to 11.0%), *p*-cymene (0 to 40.7%), 1-octen-3-ol (0.2 to 4.5%), linalool + camphor (0.9 to 67.7%), linalyl acetate (0 to 0.1%), terpinene-4-ol (0.9 to 6.5%), methyl carvacrol (1.9 to 5.2%), borneol + unknown (0.9 to 6.5%), methyl carvacrol (1.9 to 5.2%), borneol + unknown (0.9 to 18.4%), α-terpineol (1.3 to 5.5%), citronellal + carvone (0.1 to 3.7%), nerol (0 to 0.6%), geraniol (0.1 to 2.2%), cuminyl alcohol (0 to 1.0%), thymol (0 to 48.7%), and carvacrol (4.4 to 60.6%) in the various samples of *Satureia montana* oil.

Garnero et al.[236] reported that a sample of French savory oil contained α-thujene, α-pinene, camphene, β-pinene, sabinene, myrcene, α-terpinene, limonene, γ-terpinene, *p*-cymene, terpinolene, caryophyllene, β-*bis*-abolene, δ-cadinene, 1,8-cineole, 1-octen-3-ol, *trans*-sabinene hydrate, camphor, linalool, *cis*-sabinene hydrate, methyl carvacrol, terpinen-4-ol, borneol, α-terpineol, carvone, geraniol, *p*-cymen-8-ol, borneol, α-terpineol, carvone, geraniol, *p*-cymen-8-ol, cuminyl alcohol, caryophyllene oxide, thymol, and carvacrol. The authors also tentatively identified *cis*-ocimene, *trans*-ocimene, β-phellandrene, and humulene in the same oil. Chialva et al.[237] reported that an oil of *S. montana* of South American origin contained α-thujene (0.9%), β-pinene (0.7%), camphene (1.0%), β-pinene (0.2%), myrcene (1.0%), α-terpinene (1.4%), limonene (0.1%), 1,8-cineole (0.1%), γ-terpinene (6.3%), *p*-cymene (11.2%), terpinolene (0.2%), l-octen-3-ol (0.8%), linalool (0.8%), caryophyllene (3.2%), terpinene-4-ol (1.7%), α-terpineol (3.4%), β-*bis*-abolene (1.0%), caryophyllene oxide (0.9%), ethyl cinnamate (0.4%), thymol (18.4%), and carvacrol (31.3%).

Later the same authors found that the oil contained α-thujene (0.6 to 1.9%), α-pinene (0.7 to 1.5%), camphene (tr. to 0.1%), sabinene (tr. to 0.1%), β-pinene (0.2 to 0.9%), myrcene (1.8 to 2.3%), α-terpinene (2.1 to 3.5%), limonene (0.3%), 1,8-cineole (0.2%), γ-terpinene (24.6 to 32.3%), *p*-cymene (6.8 to 8.7%), terpinolene (tr. to 0.1%), 3-octanol (tr. to 0.1%), l-octen-3-ol (0.1%), *trans*-sabinene hydrate (tr. to 0.1%), l-octen-3-ol (0.1%), *trans*-sabinene hydrate (tr. to 0.1%), linalool (tr. to 0.1%), caryophyllene (0.7 to 1.8%), terpinen-4-ol-(0.3 to 0.8%), methyl chavicol (0.1 to 0.2%), humulene (tr. to 0.1%), α-terpineol (0.2%), β-*bis*-abolene (0.6 to 0.7%), *p*-cymen-8-ol (tr. to 0.1%), benzyl alcohol (tr. to 0.1%), caryophyllene oxide (0.1 to 0.6%), ethyl cinnamate (0.1%), thymol (0.2 to 0.3%), and carvacrol (44.3 to 52.5%). Trace amounts of δ-3-carene, β-phellandrene, 3-octanone, *m*-cymene, α-*p*-dimethylstyrene, furfural, camphor, β-bourbonene, terpinen-l-ol, *cis*-sabinene hydrate, bornyl acetate, methyl carvacrol, terpinen-4-yl-acetate, ethyl benzoate, piperitone, borneol, carvone, arcurcumene, geranyl acetate, δ-cadinene, thymoquinone, cuminaldehyde, myrtenol, damascenone, geraniol, phenylethyl acetate, safrole, calemen, 2-phenylethyl alcohol, cuminyl alcohol, eugenol, 2-heptadecanone, dibenzofuran, dihydroactinolide, farnesol, isoelemicin, and cumarin were also found in the same oils.

Lawrence[238] has examined three different laboratory distilled oils of *Satureia hortensis* from Europe, Canada, and North Africa. The oils showed major differences in their contents of γ-terpinene, *p*-cymene, and myrcene. The oil from Europe was very rich in γ-terpinene, to the extent of 60.3%. In addition, trace amounts of the following components were also reported:[241] *cis,cis*-allo-ocimene, 3-octanol, tetradecane, α-*p*-dimethylstyrene, pentadecane, β-ylangene, α-maaliene, calamenene, geranyl-2-methyl-propionate, piperitenone, *p*-cymen-8-ol, epi-globulol, epi-cubenol, viridiflorol, spathulenol, and 10-epi-(ol)-cadinol.

Rhyu[239] examined the comparative chemical composition of the laboratory distilled oils of two samples of savory leaves from Spain and Yugoslavia. The following are the average values in % for the Spanish and Yugoslavian oil in their percent composition: α-pinene (1.25, 1.25), camphene, (1.6, 1.3), β-pinene (0.15, 0.2), sabinene (tr.), myrcene (1.5, 0.65), α-phellandrene (1.25, 0.55), α-terpinene (0.85, 0.75), limonene (0.6, 0.4), 1,8-cineole (0.45, 0.4), γ-terpinene (3.5, 2.25), *p*-cymene (14.5, 10.7), terpinolene (0.1, 0.15), thujone (0.15, 0.7), camphor (tr.), copaene (0.15, tr.), linalool (4.5, 1.5), terpinen-4-ol (1.0, 1.0), caryophyllene (2.1, 3.7), methyl chavicol (tr.), α-terpineol (0.85, tr.), thymol (31, 13.1) and carvacrol (21., 34).

Bellomaria and Valentini[240] examined the change in chemical composition of *Satureia montana* spp. *montana* during its growth period in five different locations in the Appennino Marchigiano region of central Italy. The chemical composition of the various oils changed over the active development period of the plant. A summary of the changes in percentage composition of each component from just prior to flowering through the end of flowering to seed formation is reported by these authors.[240] Carvacrol is the major constituent followed by *p*-cymene.

G. ADULTERATION OF DRIED LEAF[1,75]

Total ash and sand content permit no conclusions on adulteration of herb by stems and petioles. High sand content indicates adulteration with earth, hence is an indication of purity. Usually ash and sand content are lower in stems and petioles than in leaf blades. It is preferable to calculate crude fiber, crude fixed oil, crude protein, and total ash on a water-free and sand-free basis. Increased crude fiber content occurs in adulteration by stems and petioles whereas in such cases crude fixed oil, protein, and ash are lower. These herbs are best washed before drying and milling, by which means the sand content may be reduced by 10%.

H. SPECIFICATIONS[57]

The standard specifications for the oil of savory as given by the essential oil association of the U.S. are given in Table 52.

I. USES[1,3,5,6,12,60,70,74,75]

The fresh, tender young leaves and tips may be minced and used to flavor egg dishes, meats, and poultry dressings. Fish salads, stems, sauces, and vegetables take a new flavor when the tiny leaves of summer savory are used instead of parsley or chives. One teaspoon of ground savory equals 0.053 ounces. In general, the spice or the oil of summer savory is used in the manufacture of liquors, bitters, and vermouths, some baked goods, condiments, and in very small proportions in confectionaries. It is also used in commercial dry soup mixes and gravy mixes and is seldom used in seasonings.

It is used in medicines as a carminative and expectorant but its importance today is almost entirely culinary. The herb has an antioxidant property.

TABLE 52
E.O.A. Standards (1966) for Oil from Savory (Summer Variety)[57]

Characteristics	Specifications
Physical and chemical constants	
Appearance and odor	Light yellow to dark brown liquid having a spicy aromatic note reminiscent of thyme or organism
Specific gravity at 25/25°	0.875—0.954 (temperature correction factor 0.00065/°C)
Optical rotation	−5°—+4°
Refractive index at 20°C	1.4860—1.5050
Phenol content (carvacrol)	20—57%
Saponification value	Up to 6.0
Solubility in alcohol at 25°C	Generally soluble in 2.0 vols of 80% alcohol; occasionally some oils are slightly hazy in 10 vols of 90% alcohol
Descriptive characteristics	
Solubility	
Benzyl benzoate	Soluble in all proportions
Diethyl phthalate	Soluble in all proportions
Fixed oils	Soluble in all proportions in most fixed oils
Glycerine	Relatively insoluble
Mineral oil	Soluble in all proportions
Propylene glycol	Relatively insoluble
Stability	
Alkali	Not stable in the presence of alkali
Acid	Unstable to strong inorganic acids but stable to weak organic acids
Containers	Can be shipped in glass, aluminum, tin-lined, or other suitably lined containers
Storage	Store in tight, full containers in a cool place protected from light

II. *SATUREIA MONTANA* L. — (WINTER SAVORY)[70]

Savory or winter savory is another popular species of the more than half a dozen dwarf or bushy savories. Like the summer savory it is native to the Mediterranean areas and is also grown extensively in all parts of the U.S. and Europe, but is said to have less flavor attributes to it as compared to summer savory.

A. BOTANY[64]

The winter variety is a beautiful, hardy perennial with dark green, shiny leaves which have a slightly stronger and more pungent odor and flavor than that of the summer savory. Its lonely pinkish white and sometimes almost purple flavors resemble the tiny blossoms of the mint, mentha family. The flowers bloom continuously from mid-summer until frost.

B. CHEMICAL COMPOSITION[1,75]

The physical properties of the following results are reported:[13,75] yield 0.18%, specific gravity 0.939; optical rotation 2°35′, and solubility in 1.5 vols of 80% alcohol, and 65% phenol. Guenther[13] reports a yield 0.1 to 0.2% of oil, showing specific gravity at 15°C, 0.924; refractive index at 20°C 1.4918; optical rotation +2.8°; ester number 7.5; and solubility in 2.1 vols of 80% alcohol. In addition to terpenes, 32% of carvacrol is present as described earlier.

C. USES[3,16,69,70,74,75]

Also like the summer variety, the winter variety is used profusely. The leaves give an added taste to poultry stuffings and some egg dishes. It may be used instead of parsley or blended with it and other herbs. Though not quite so delicate in flavor as the summer savory, it may also be used in place of it and in a bouquet of garni for soups and stems. It is also reported to be carminative and a stimulant.

Chapter 20

TARRAGON

I. *ARTEMISIA DRACUNCULUS* (L.) — FAMILY: COMPOSITAE

Tarragon, is a small herbaceous perennial plant of the compositae family indigenous to southern Russian and western Asia. It is believed to be indigenous to southeastern Russia but it is now cultivated throughout Europe and western Asia and France. The U.S. grows sufficient quantities for its own needs. It is also distributed in western Tibet and Afghanistan.[3]

A. BOTANY AND CULTIVATION[3,60,74,75]

The tarragon plant is believed to be a native of Siberia. There are two varieties of this species of *Artemisia*: one known as Russian Tarragon, and the other is the variety cultivated in the U.S. The French tarragon plant has bright green, lance-shaped leaves. The flowers are not always produced and are small very pale yellow in color, and generally sterile. French tarragon cultivated in the U.S. produces no flowers. These plants are propagated by root divisions and cuttings.

Tarragon grows well in most temperate climates, although it does best in warm, dry, well-drained light soils in sunny locations. Two or three crops of the young herbage may be harvested annually.

The leaves are linear-pointed, sestile, and hardy reaching 8 cm in length and 1 cm in width. The midrib and two ribs, one on each side running along the margin from base to tip are conspicuous, even in young leaves in which the veins are indistinct. As regards the microscopic structure of the leaf, both epidermis are much alike, the cells over the veins being elongated, straight-walled, faintly beaded, and striated while between the veins they are sinnuate-walled, especially on the lower epiderm.[75] Stomata and short capitate hairs, with short stalk and two-celled head occur occasionally. Palisade cells occur on both sides of the leaf, usually in two rows, one on the upper side and one on the lower; through the remainder of the mesophyl a thin layer runs through the fibrovascular bundles. Accompanying each bundle of the midrib on the upper side is a large oleoresin duct like those of the stem.[75]

When the plants are well established, the green leaves and tender top growth may be harvested at intervals during the growing season, by cutting the stems a few inches above ground level.[75] The freshly cut crop is shade dried to preserve the leaf color and is then stored in closed containers to retain as much as possible of the warmly aromatic licorice-anise aroma. The leaves are also dried indoors by artificial heat and circulating air, carefully controlled to ensure retention of aroma and color.[3,75] The dried leaves are then processed through a series of machines to provide particle sizes suitable to the needs of consumers.[74,75]

B. COMPOSITION[72]

The composition of dried herb is shown in Table 53.

C. SPECIFICATIONS IN THE U.S.[5]

The dried leaves should be from the perennial herb of *Artemesia dracunculus* L. The light to dark green leaves have a pleasant, aromatic, licorice-anise-like odor and possess a bittersweet and herbaceous odor. The dried product should not contain more than 15% ash, 1.5% total insoluble ash, 10% moisture, and not less than 1.3 ml of volatile oil per 100 g material. The dried powder should pass through a U.S. standard No. 40 sieve.

TABLE 53
Composition of Ground Tarragon (Dry)
(100 g Edible Portion)

Component	Quantity
Water	4.7—8.0 g
Food energy	295—365 kcal
Protein	22.8—25 g
Fat	7—7.5 g
Total carbohydrate	45—50 g
Ash	12.0 g
Fiber	7.4 g
Calcium	1140 mg
Iron	32 mg
Magnesium	347 mg
Phosphorus	313 mg
Potassium	3020 mg
Sodium	62 mg
Zinc	4 mg
Niacin	9 mg
Riboflavin	1 mg
Vitamin A (as β-carotene)	4200 I.U.
Vitamin C	12 mg

TABLE 54
Physicochemical Properties of Tarragon (*Artemisia dracunculus*) Oil[75]

Characteristics	Volatile oil		Terpene-less oil
	Minimum	Maximum	
Specific gravity at 15°C	0.900	0.981	0.975
Refractive index at 20°C	1.5028	1.5160	1.5206
Optical rotation	+2°	+9°	—
Ester number	1	9	—
Acetyl number	15	22	—
Acid number	0	1	—
Solubility in 80% alcohol (vols)	6	11	3

D. VOLATILE OIL

Steam distillation of the herb yields about 0.3 to 1.3% of a pale yellow essential oil with an anise-like odor. On an average, the oil content varies from 1 to 2% in most of the varieties.[3]

Analysis of the oil cited by Guenther[13] indicated a lower refractive index (1.4712 to 1.490) and a higher rotation (+20 to +30); the ester number varies from 7.4 to 13.99 and the ester number after acetylation is 30 to 42 as compared to the values reported in Table 54.

Thieme and Tarn[242] identified the following in tarragon oil: α-pinene, camphene, β-pinene, limonene, *cis*-ocimene, *trans*-ocimene, methylchavicol, *p*-methoxy cinnamaldehyde, -4-carene, α-phellandrene, and linalool. More recently, Zarghami and Russell[243] found that the oil contained limonene, *cis*- and *trans*-ocimene, *cis*- and *trans*-allocimene, linalool, butyric acid, methyl chavicol,(major), geraniol, 1,2-dimethoxy li-allyl benzene, and eugenol.

Vostrowsky et al.[244] examined the chemical composition of the steam distilled fresh leaves of tarragon using GC-MS. The results of their study can be summarized as follows: α-thujene (0.16%), α-pinene (0.52%), camphene (0.03%), sabinene (38.81%), β-pinene

TABLE 55
E.O.A. Standard (1966) Specifications for the Oil of Tarragon (*Artemisia dracunculus*)[57]

Characteristics	Standards
Preparation	Obtained through steam distillation of the leaves, stems, and flowers of the plant
Appearance and odor	Pale yellow to amber liquid, having a delicate spicy odor reminiscent of licorice and sweet basil but characteristic of estragon oil
Physical and chemical constants	
Specific gravity at 25°C	0.914—0.956
Optical rotation	+1°30′—6°30′
Refractive index	1.5040—1.5200
Acid value	Max. 2.0
Saponification value	Up to 18
Solubility in alcohol	0.5 to 1.0 and more vols of 90% alcohol
Solubility	
Benzyl benzoate	Soluble in all proportions
Diethyl phthalate	Soluble in all proportions
Fixed oils	Soluble in all proportions
Glycerine	Insoluble
Mineral oil	Soluble in equal vols
Propylene glycol	Relatively insoluble
Stability	
Alkali	Unstable for strong alkalies
Acid	Unstable in the presence of mineral acid
Containers	Preferably in glass or lined containers
Storage	Store in tight full containers in a cool place protected from light

(0.17%), myrcene (1.62%), α-terpinene (0.56%), *p*-cymene (0.03%), limonene (0.75%), (Z)-ocimene (2.06%), (E)-ocimene (1.97%), γ-terpinene (0.94%), *trans*-sabinene hydrate (0.07%), terpinolene (0.26%), *cis*-sabinene hydrate + linalool (0.13%), terpinen-4-ol (1.16%), methyl chavicol (17.26%), α-terpineol (0.04%), bornyl acetate (0.05%), citronellyl acetate (0.89%), geranyl acetate (0.45%), methyl eugenol (28.87%), cinnamylacetate (0.14%), methyl isoeugenol (1.15%), elemicin (0.34%), farnesene (0.07%).

Later Brass et al.[245] characterized the presence of elemicin, *trans*-isoelemicin, eugenol, methyl eugenol, and *trans*-methyl isoeugenol in an oil of the Russian tarragon. Subsequently, Albasini et al.[246] compared the chemical composition of lab-distilled tarragon oils obtained from the French, German, and Russian tarragon cultivars. Methyl chavicol (73%) was the main constituent followed by *cis*- and *trans*-ocimenes and limonene.

Later Bayrak and Dogan[247] examined the chemical composition of Turkish tarragon oil produced from plants harvested in the vicinities of Erzurum and Gaziantep respectively. He also indicated methyl chavicol (74.2%) as the main constituent along with anisaldehyde (1.1%).

Lawrence[248] has examined the chemical composition of several forms of tarragon; the results indicate that similar to earlier results of Brass et al.[245] and Bayrak and Dogan[247] methyl chavicol (73%) was the main ingredient.

E. SPECIFICATIONS[57]

The standard specifications for the essential oil of tarragon as specified by the E.O.A. are shown in Table 55.

F. USES[1-6,12,60,70,74,75]

Both the fresh and the dried leaves are used to flavor many foods. Minced or chopped finely and spread over steaks, salads, fish, and egg dishes, the herb lends a piquancy to the flavor that is unusually delicious and aromatic. Bernaisse, Hollandaise, and tartar sauces are often flavored with tarragon. Marinades for fish and meats are improved by the judicious use of this herb.[74] Tarragon blends well with tomato juice, 'fines herbs', and fish sauces or may be sprinkled on fresh salads, meats, and stems. Its flavor is so pungently distinctive however, that it is best to use it sparingly.[75] Tarragon vinegar, a favorite of culinary preparations is prepared by saturating the fresh or dried herb in wine vinegar. Tarragon vinegar has many uses especially in salad dressings and mayonnaise. It is also equally important in mixed mustards.[3,6]

Dried tarragon sometimes acquires a hay-like overtone and hence many culinary preparations require fresh tarragon. The leaf is widely used with chicken and egg preparations. As it is used mainly in delicate dishes, it must be added with good judgment. Tarragon also flavors various liquors and is used in perfumery.[3] The leaves are credited with aperient, stomachic, stimulant, and febrifuge properties.

Another plant, *Tagetes lucida* has a flower and aroma close to that of tarragon and many times is used as a substitute.[3]

Chapter 21

THYME

I. *THYMUS VULGARIS* (L.) — FAMILY: LABIATAE

Thyme consists of the dried leaves and flowering tops of *Thymus vulgaris* L. a perennial shrub belonging to the Labiatae, (mint) family. Thyme is native to southern Europe and the Mediterranean region. It is cultivated in Spain, Portugal, France, Germany, Italy, and other countries of continental Europe, in North Africa, England, Canada, and the U.S. Spain and France are the leading suppliers of thyme for the U.S. market; a small quantity comes from Portugal and the balance is produced by American growers. World export of thyme is about 1500 tonnes.[3] It grows wild in the dry hills of southern Europe. It is a common garden plant which lives for many years under good culture. Much of today's commercial production is in southern Europe including France, Spain, Portugal, and Greece.

A. BOTANY AND CULTIVATION[3,57,64]

Thyme is a diminutive perennial herbaceous shrub of the mint family native to the Mediterranean region and Asia Minor. *Thymus vulgaris* of garden type is the kind most frequently found in herb gardens. It is suberect with numerous stems 8 to 18 in. high and has a woody fibrous root. The tiny grayish-green narrow leaves rarely exceed 1/4 in. in length and by about 1/10 in. in width.

Among the other fragrant useful thymes are the creeping ('mother of thyme') which grows in thick mats, the lemon, the mint, the orange, the golden lemon, etc. The leaves vary in form from broad-leaved to narrow-leaved. These are upright as well as prostrate trailing species.[3]

Thyme prefers a mild climate and well-drained sunny location. A spacing of 3 ft by 1 ft should be left between the plants. Thorough weeding is essential. When growing wild in very dry conditions, the leaves are exceedingly sparse. Thyme is easy to grow from seeds or from pieces of plant with a bit of root attached. It prefers sun and dry conditions in light calcareous soil. In damp clay it can be difficult and will not have such a good aroma. Thyme is perennial, but some people advocate replanting every few years.[5]

The yield of herb is 1000 to 2000 lb/acre; with good irrigation the yield can be increased. The flavor of thyme is distinctively warm, pleasant, aromatic, pungent, and reminiscent of sage with lingering sharpness.[64]

B. EXTRACTIVES[5]

The volatile oil of thyme is a pale, yellowish-red liquid with a rich sweet, aromatic herbaceous odor yielding a sweet, phenolic, somewhat medicinal perception upon drying. The taste is sharp, biting, warm, spicy, and herbaceous.

The oleoresin of thyme is a dark green to brown, somewhat viscous, at times almost a semisolid, with a minimum volatile oil content of 50 ml/100 g. Generally, 1.81 kg (4 lb) of oleoresin is equivalent to 45.45 kg (100 lb) of freshly ground, dried thyme in flavor and aroma attributes.[5]

C. SPECIFICATIONS IN THE U.S.[5]

The spice is the dried leaves and the flowering tops of *Thymus vulgaris* L. The dried brownish-green curled leaves, when crushed should yield a fragrant, aromatic odor and have a warm, pungent taste. It shall contain not more than 11.0% total ash, 5.0% acid insoluble ash, and 9.0% moisture, and no less than 0.9 ml of volatile oil (vol wt) per 100 g. Not less than 95% passes through a U.S. standard No. 30 sieve.[5]

TABLE 56
Composition of Ground Thyme (100 g Edible Portion)[5]

Component	Quantity
Water	7.8 g
Food energy	257—350 kcal
Protein	6.8—9.1 g
Fat	4.6—7.4 g
Total carbohydrate	48—63.9 g
Pentosans	12—16 g
Fiber	19—24.0 g
Ash	11.7—13.2 g
Calcium	1890 mg
Iron	124 mg
Magnesium	220 mg
Phosphorus	201 mg
Potassium	814 mg
Sodium	55 mg
Zinc	6 mg
Niacin	5 mg
Vitamin A (as β-carotene)	3800 I.U.

TABLE 57
The Physicochemical Properties of Oil of Thyme (*Thymus vulgaris* L.)[13]

Constants	Spanish	Moroccan	Sardinian
Specific gravity at 25/25°	0.916—0.934	0.891—0.910	0.902—0.904
Optical rotation at 25°	−0°16′—1°52′	−2°20′— −3°12′	—
Refractive index at 20°	1.4971—1.5040	1.4909—1.4967	1.489—1.497
Phenol content	42.5—59.0%	28—37%	44—54%
Solubility at 25°	2.5—3.5 vols and more of 70% alcohol and 3 vols and more of 80% alcohol	1—1.5 vols and more of 80% alcohol	1—1.5 vol% at 20°

D. COMPOSITION

Thyme herb dried, has the composition as shown in Table 56.[5]

E. VOLATILE OIL[13,55,249,250]

Usually a yield of 2 to 2.5% of the oil is obtained. It is an odorless to yellowish-red liquid with a pleasant odor characteristic of the herb. The ordinary thyme oil contains 42 to 60% phenols mainly as thymol (crystallizable). The lemon type thyme oil contains citral which accounts for some the lemon-like odor.[13,74]

The white thyme oil will sometimes be yellowish-red to deep red in color. It is mainly due to the action of phenol on iron (from condensers in the field stills). A colorless oil rich in phenols can be had by redistilling. Moroccan thyme will yield somewhere between 0.5 to 1.2% of oil.[250]

The physicochemical properties of the different types of thyme oil are presented in Table 57.

The oil contains thymol (40 to 60%), carvacrol, geraniol, amyl alcohol, *l*-borneol, *l*-linalool, β-hexenol, 1-α-pinene, β-pinene, camphene, *p*-cymene, γ-terpinene, caryophyllene, α-pinene, myrcene, 1,8-cineole, 3-carene, sabinene, camphene, α-thujone, terpinolene, ocimene, caffeic acid (1.84%), chlorogenic acid (1.13%), citral, cymol, terpineol, and camphene.[3,249,250] By using gas chromatography and infrared spectroscopy it has been demonstrated that γ-terpinene isolated from *Thymus vulgaris* is transformed spontaneously in air to *p*-cymene. It has been established that in most of the samples, pinenes and traces of borneol were present.[249] The Raman frequencies recorded for this oil agree well with those of α-pinene, the main constituent of the oil, β-pinene, and limonene. The frequencies at 710 and 750 cm^{-1} may be attributed to the presence of β-pinene. While those at 1537, 1600, 1680 may be due to the presence of limonene.[249]

Thymol, methylisopropylphenol or oxyhydroxycymene, is a well known and efficient antiseptic much used in mouthwashes, etc. The substance may be prepared from the oil of thyme or oil of ajowan or may be synthesized from dibromomenthone. The colorless hexagonal crystalline plates melt at 50°C and boil at 230°C. They are soluble in alcohol and ether, but only slightly in water.

Carvacrol or isoprophyl-*o*-cresol, $C_{10}H_{14}O$, or $C_6H_3(CH_3)(OH)$ (C_3H_7) is an isomer of thymol and occurs with it in oil of thyme. Savory and spearmint oils also contain carvacrol. It may be separated from thyme oil or prepared by synthesis, as a thick, oily, optically inactive substance having specific gravity at 15°, 0.981; and boiling point 237°C.

Ikeda et al.[79] identified the following in the monoterpene fraction (29.7%) of thyme oil: α-pinene (3.7%), α-thujene (2.3%), camphene (1.5%), β-pinene (1.2%), δ-3-carene (0.2%), α-phellandrene (0.4%), α-terpinene (4.4%), myrcene (6.3%), limonene (1.3%), γ-terpinene (19.4%), and *p*-cymene (59.0%). Masada[173] used a combination of GC and MS to characterize the presence of α-pinene, camphene, β-pinene, myrcene, α-terpinene, limonene, 1,8-cineole, γ-terpinene, *p*-cymene, linalool, terpinen-4-ol, caryophyllene, bornyl acetate α-terpineol, borneol, thymol (major constituent), and carvacrol in an oil of red thyme. Later Rhyu[239] identified α-pinene (1.0 to 3.7%), camphene (0.7 to 2.4%), β-pinene (tr. to 0.6%), sabinene (tr. to 0.5%), myrcene (tr. to 2.6%), α-phellandrene (tr. to 0.3%), α-terpinene (tr. to 0.2%), limonene (0.4 to 2.1%), 1,8-cineole (0.4 to 7.4%), γ-terpinene (0.3 to 6.5%), *p*-cymene (7.3 to 42.5%), terpinolene (tr. to 2.0%), thujone (0.1 to 0.8%), camphor (tr. to 0.2%), copaene (tr. to 0.2%), linalool (1.3 to 12.4%), terpinen-4-ol (0.3 to 9.5%), caryophyllene (tr. to 1.0%), methyl carvacrol (tr. to 0.1%), α-terpineol (0.4 to 0.9%), thymol (7.7 to 39.9%), and carvacrol (0.5 to 4.8%).

Subsequently Mateo et al.[251] examined the chemical composition of *Thymus vulgaris* oil along with five other samples of *Thymus zygis* oils of Spanish origin. The average results of the various constituents of *Thymus vulgaris* and *Thymus zygis* respectively in parenthesis are α-pinene (5.1, 3.6), camphene (11.4, 4.2), β-pinene (2.8, 1.2), sabinene (0.6, 0.5), myrcene (5.4, 3.0), limonene (3.0, 1.0), 1,8-cineole (33, 7.1), γ-terpinene (6.1, 8.0), *p*-cymene (6.8, 15.1), camphor (14.5, 11.5), linalool (0.9, 3.5), linalyl acetate (0, 0.45), caryophyllene (0.9, 1.1), α-terpineol (1.5, 1.5), borneol (3.5, 3.6), alloaromadendrene (0, 0.1), terpinen-4-ol (2.4, 0.85), methoxy carvacrol (traces), thymol (tr., 35.6), and carvacrol (0, 10.5).

Lawrence[252,254] reported that an oil of thyme was found to contain α-thujene + α-pinene (4.5%), camphene (2.0%), β-pinene (0.5%), sabinene (0.4%), verbenene (0.5%), myrcene (1.2%), α-phellandrene (0.1%), α-terpinene (1.4%), limonene (1.1%), 1,8-cineole (5.6%), *trans*-ocimene, γ-terpinene + 3-octanone (5.6%), *p*-cymene (22.8%), l-octen-3-ol (0.3%), α-thujone + α-*p*-dimethylstyrene (0.2%), *trans*-linalool oxide (0.1%), camphor + *cis*-sabinene hydrate (0.1%), linalool + β-bourbonene (8.0%), linalyl acetate + *cis*-*p*-menth-2-en-l-ol (2.2%), bornyl acetate (0.5%), methyl thymol, methyl carvacrol + terpinen-4-ol (4.5%), caryophyllene (1.8%), alloaromadendrene (0.3%), isopinocarveol (0.2%), borneol (0.5%), α-terpineol (4.2%), α-terpinyl acetate (5.7%), verbenone (3.9%), geranyl acetate (0.2%), *trans*-anethole (0.1%), *p*-cymen-8-ol (0.5%), caryophyllene (0.1%), thymol (15.9%), carvacrol (4.3%). In addition, trace amounts of 2-ethylfuran, methyl 2-methylbutyrate, toluene, terpinolene, 3-octanol, α-cubebene, copaene, α-gurjunene, α-maaliene, aromadendrene, γ-muurolene, α-muurolene, α-selinene, δ-cadinene, γ-cadinene, geraniol, α-cadinene, calamenene, α-calacorene were also identified in the same oil.

Passel[253] reported that the thyme populations across the southern Mediterranean regions (collected in the wild) were found to contain essential oils with differing chemical compositions. For example, Passel found the following composition: thymol/carvacrol (0.2 to 72%), *p*-cymene (2 to 40%), geraniol (0.2 to 41%), linalool (0.2 to 75%), *cis*-myrcen-8-ol (0.2 to 25%), 1,8-cineole (tr. to 70%), α-terpineol (1 to 65%), terpinene-4-ol (1 to 30%), *trans*-sabinene hydrate (0.5 to 42%).

1. Wild Thyme Oil

Agarwal and Mathela[255] reported that an oil obtained from wild thyme (*Thymus serpyllum*) was found to contain α-pinene (0.7%), camphene (0.6%), β-pinene (0.3%), β-carene (1.1%), α-terpinene (0.4%), limonene (0.2%), γ-terpinene (6.1%), *p*-cymene (13.5%), terpinolene (tr.), 1,8-cineole (0.2%), citronellal (trace), linalool (0.1%), *trans*-β-terpineol (4.2%), terpinene-4-ol (2.6%), geraniol (5.9%), and thymol and carvacrol (48% total phenols). The authors used GC retention time data and TLC as their method of identification. Later, Agarwal and Mathela[256] re-examined the chemical composition of the same oil. Using GC-MS as the method of analysis, the oil was found to contain α-thujene, α-pinene, camphene, β-pinene, sabinene, 3-carene, myrcene, α-terpinene, limonene, β-phellandrene, 1,8-cineole, γ-terpinene, 3-octanone, *p*-cymene, terpinolene, 3-nonarone, α-*p*-dimethylstyrene, *trans*-sabinene hydrate, camphor, *cis*-sabinene hydrate, bornyl acetate, caryophyllene, methyl thymol, α-humulene, borneol, β-*bis*-abolene, δ-cadinene, thymyl acetate, carvacryl acetate, thymol (60%), and carvacrol (2%). It is worth noting that terpinen-4-ol and *trans*-β-terpineol, which were previously characterized by retention time data, were not identified by mass spectrometry.

Lawrence[257] reported the results of four analyses of *T. serpyllum* oils. His results indicate carvacrol being the main component (37.1%) followed by γ-terpinene and *p*-cymene.

F. ADULTERATION[13]

Thyme oil is usually adulterated by the addition of terpenes (*p*-cymene-camphene + γ-terpinene). The phenol content of the oil will be decreased correspondingly. If synthetic carvacrol is used it can be detected by non-crystallizable phenols.

G. USES[1-6,12,16,52,60,70,71,74,75,249]

The dried or fresh flowers and leaves are used in many foods and the dried, brownish-green leaves are often ground and blended with other herbs to be used in special stuffings and seasonings. In Switzerland, thyme is used to flavor a special creamy goat's milk cheese and the famous Benedictine liquor is flavored with thyme. Creams, custards, croquettes, vegetable cocktails, fish, shellfish, meat stuffings, chowders, and soups are all improved by the use of this warm, pungent herb. Salads and sauces owe much to thyme; and butters, jellies, and vinegars are intriguing and delicious when flavored with it. It is also used in preparing herb teas and herb vinegars. It finds use in pickles to flavor olives.[3]

Thymol from the oil is utilized in pharmaceutical preparations, such as gargles, cough-drops, dentrifices and mouthwashes. It is also used to preserve meat and as a vermifuge to cure hookworm in horses and dogs.[3,69,71]

In herbal medicine thyme cures nightmares and melancholy and aids menstrual flow. It is reported to be good for strengthening the lungs and improving digestion.[71]

REFERENCES

1. **Parry, J. W.,** *'Spices', Morphology, Histology and Chemistry,* Vol. 2, Chemical Publishing Co. Inc., New York, 1969.
2. **Svendsen, A. B. and Scheffer, J. J. C., Eds.,** Essential Oils and Aromatic Plants, Proc. 15th Int. Symp. on Essential Oils, Noordwijkerhout, The Netherlands, July 19—21, 1984.
3. **Prakash, V.,** Chemical Composition and Uses of Leafy Spices, Dissertation submitted for partial fulfillment of M.Sc. degree, University of Mysore, Mysore, India, 1972.
4. **Lewis, Y.S.,** *Spices and Herbs for the Food Industry,* Food Trade Press, Orpington, England, 1984.
5. **Farrell, K.,** *Spices, Condiments, and Seasonings,* AVI Publ., Westport, CT., 1985.
6. **Pruthi, J. S.,** *Spices and Condiments,* National Book Trust, New Delhi, 1976.
7. **Heath, H. B.,** *Natural Food Flavorings,* Symposium on Flavourings in Food, British Society of Flavorists, 1972.
8. **Anon.,** *Spices — A Survey of the World Market,* Vol. 2, International Trade Centre UNCTAD/GATT, Geneva, 1977.
9. **Anon.,** Proceedings of Symposium on Spice Industry in India, Association of Food Scientists and Technologists, Mysore, India, 1974.
10. **Furia, T. F. and Bellanca, Eds.,** *Fenaroli's Handbook of Flavor Ingredients,* Vol. 2, CRC Press, Cleveland, OH, 1975.
11. **Chopra, R. N., Nayar, S. L., and Chopra, I. C.,** *Glossary of Medicinal Plants of India,* CSIR, New Delhi, 1956.
12. **Council of Scientific and Industrial Research (CSIR),** *Wealth of India — Raw Materials,* Vols. I — X, Publication Directorate, CSIR, New Delhi, India, 1949—1975.
13. **Guenther, E.,** *The Essential Oils,* Vols. 1—6, D. Van Nostrand, New York, 1949.
14. **Salzer, U. J.,** Analytical evaluation of seasoning extracts (oleoresins) and essential oils from seasonings, *Int. Flavors Food Additives,* 6, 151—157, 206—210, 253—268, 1975.
15. **Mukerji, B.,** The Indian Pharmaceutical Codex, CSIR, New Delhi, 1953.
16. **Hemphill, R.,** *The Penguin Book of Herbs and Spices,* Penguin Books, Ltd., Harmondsworth, Middlesex, England, 1966.
17. **Winton, A. L. and Winton, K. B.,** *Structure and Composition of Foods,* Vol. IV, John Wiley & Sons, New York, 1939.
18. **Gopalan, C., Ramasastri, B. V., and Balasubramanian, S. C.,** *Nutritive Value of Indian Foods,* National Institute of Nutrition, Indian Council of Medical Research, Hyderabad, India, 1978.
19. **American Spice Trade Association,** *Your Spice Shelf Cookbook,* Prentice Hall, Englewood Cliffs, NJ, 1972.
20. **Anon,** *''When is Herb a Spice'',* American Spice Trade Association, Englewood Cliffs, New Jersey, 1980.
21. **Jacobs, M. B.,** *The Chemistry and Technology of Foods and Food Products,* Vols. I and 2, Interscience Publ., New York, 1944.
22. **Cartwright, I. C. and Nanz, R. A.,** Comparative evaluation of spices, *Food Technol.,* 2, 330—336, 1948.
23. **Central Food Technological Research Institute,** A Select Bibliography on Spices (1966—1972), Mysore, India, 1974.
24. **Claibourne, C.,** *Cooking with Herbs and Spices,* revised ed., Harper & Row, New York, 1970.
25. **Collins, R. P.,** *A World of Curries — An International Cookbook,* Harper & Row, New York, 1967.
26. **Desrosier, N. W.,** *Rietz Master Food Guide,* AVI Publ., Westport, CT, 1978.
27. **Eiserle, R. J. and Rogers, J. A.,** The composition of volatile oils derived from oleoresins, *J. Am. Oil Chem. Soc.,* 49, 573—577, 1972.
28. *Encyclopedia of Herbs,* Marshall Cavendish Co., London, 1977.
29. **FDA,** 19 Spices Defined, 21 — Code of Federal Regulations 1.12(a), U.S. Food and Drug Administration, Washington, D.C.
30. Spices, essential oils, oleoresins and natural extractives generally recognized as safe, *Fed. Reg.,* 26, 5221, June 10, 1961.
31. **Lawrence, B. M.,** *Perfumer Flavorist,* 7, 47—48, 1982.
32. Federal Specification, Spices, Ground and Whole, and Spice Blends, EE-S-631H, Washington, D.C., June 5, 1975.
33. **Heath, H.,** Herbs and spices — a bibliography, Parts I and II, *Flavor Industry,* 4, 24—26, 65—66, 1973.
34. **Gaulke, J. A.,** *Cooking with Spices and Herbs,* Lane Magazine and Book Co., Menlo Park, CA, 1975.
35. **Gerard, J.,** *The Herbal and General History of Plants,* Dover Publ., New York, 1975.
36. Good Housekeeping Institute, *Good Housekeeping Cooking with Herbs and Spices,* Ebury Press, London, 1975.
37. **Green, R. J.,** Peppermint and spearmint production in the United States: progress and problems, *Int. Flavors Food Additives,* 6, 246—247, 1975.

38. **Grieve, M.,** *A Modern Herbal,* The Medicinal Culinary, Cosmetic and Economic Properties, Cultivation and Folklore of Herbs, Grasses, Fungi, Shrubs and Trees, with All their Modern Scientific Uses, Hafner Press, New York, 1974.
39. **Hardman, R.,** *Spices and Herbs; their Families, Secretory Tissues and Pharmaceutical Aspects,* Tropical Products Institute, Conference Proceedings, London, 1973.
40. **Heath, H. B.,** *Flavor Technology: Profiles, Products, Applications,* AVI Publ., Westport, CT, 1978.
41. **Heath, H. B.,** Herbs and spices for food manufacture, *Trop. Sci.,* 14, 245, 1978.
42. **Heath, H. B.,** Herbs and spices — a Bibliography, Part I, *Flavor Ind.,* 4, 24—26, 1973.
43. **Hemphill, R.,** *Cooking with Herbs and Spices,* Angus and Robertson, London, 1977.
44. **Hodgson, M.,** *The Hot and Spicy Cookbook,* McGraw-Hill, New York, 1977.
45. **Julseth, R. M. and Diebel, R. H.,** Microbial profile of selected spices and herbs at import, *Indian Spices,* 11, 6—11, 1974b.
46. **Kreuger, J.,** Dehydrated ingredients: availability, advantages, disadvantages, use and cost, *Proc. Prep. Foods.,* 3, 1980.
47. **Law, D.,** *The Concise Herbal Encyclopedia,* St. Martin's Press, New York, 1973.
48. **Lowefeld, C.,** *The Complete Book of Herbs and Spices,* 2nd ed. revised, David and Charles Publ., New York, 1978.
49. **Menuti, W.,** *Freeze-Dried Chopped Chives,* G. Armamino and Son, Inc., San Francisco, CA, 1980.
50. **Russell, G. F. and Olson, K. V.,** The volatile constituents of oil of thyme, *J. Food Sci.,* 37, 405—407, 1972.
51. **Stobart, T.,** *The International Wine and Food Society's Guide to Herbs, Spices and Flavorings,* McGraw-Hill, New York, 1973.
52. **Heath, A.,** *Herbs in the Kitchen,* Faver and Faber Co., London, 1953, 15.
53. **Baslas, K. K.,** Chemistry of Indian essential oils, *Perfumery and Essential Oil Rec.,* 59, 103—109, 1968.
54. **Singh, H. S., Bhagat, S. D., Mathur, R. K., Gopinath, K. W., and Ganguly, D.,** Cultivation of basil, *Ocimum basilicum,* L. at Jorhat, Assam and the chemical composition of its oil, *Flavor Ind.,* 2, 481, 1971.
55. **Pearson, O.,** *The Chemical Analysis of Foods,* J. & A. Churchill, London, 1970.
56. **Guenther, E.,** *The Essential Oils,* Vol. 3, Van Nostrand, New York, 1949, 395—433, 519—761.
57. **E.O.A.,** Book of Standards and Specifications, Essential Oil Association of U.S.A., New York, 1966.
58. **Lawrence, B. M., Terhune, S. J., Hogg, J. W.,** *Flavor Industry,* 2, 173—175, 1971.
59. **Menon, A. K.,** *Indian Essential Oils. A Review,* Council of Scientific and Industrial Research, New Delhi, 1960.
60. **Rosengartner, F., Jr.,** *The Book of Spice,* Churchill Livingstone, London, 1969.
61. **Ivanov, D., Tchorbad, J. S., Iordanov, Tch.,** On the composition of Bulgarian basil oil, *Perfumery Essential Oil Rec.,* 55, 717—719, 1964.
62. **Kirtikar, K. R. and Basu, D. D.,** *Indian Medicinal Plants,* Vol. 1, Lalit Mohan Basu, Allahabad, India, 1959—1968.
63. **Council of Scientific and Industrial Research (CSIR),** *Wealth of India — Raw Materials,* Vol. 7, Publication Directorate CSIR, New Delhi, 1966, 79—89.
64. **Sobti, S. N., Pushpangadan, P., and Atal, C. K.,** *Genus ocimum* — a potential source of new essential oils, *Indian Perfumer,* 20, 59—68, 1976.
65. **Gupta, S., Thapa, R. K., Vashist, V. N., Madan, C. L., and Atal, C. K.,** Introduction of French Basil, *Ocimum basilicum,* in Jammu, cultural practices and chemical constituents, *Flavor Ind.,* 2, 707—708, 1971.
66. **Brophy, J. and Jogia, M. K.,** Essential oils from two varieties of Fijian *Ocimum sanctum* (tulsi), *Fiji Agric. J.,* 46, 21—62, 1984; *Chem. Abstr.,* 104, 10350u, 1986.
67. **Kirk-Othmer,** *Encyclopedia of Chemical Technology,* Vol. 9, 2nd ed., Interscience, New York, 1966.
68. **Dejoic, L.,** Oil of Haitian sweet basil, *Drug Cosmet. Ind.,* 53, 510—511, 579—581, 1943; *Chem. Abstr.,* 38, 16068, 1946.
69. **Quelch, M. T.,** *Herbs for Daily Use in Home Medicine and Cookery,* Faber and Faber Ltd., London, 1961, 34.
70. **Miloradovich, M.,** *The Art of Cooking with Herbs and Spices,* Doubleday and Co., New York, 1952.
71. **Nadkarni, A. K.,** *Indian Mater. Med.,* Popular Book Depot, Bombay, 1954.
72. **Parry, E. J.,** *The Chemistry of Essential Oils and Artificial Perfumes,* Vol. I, 4th ed., Scott Greenwood and Son, London, 1971.
73. Spices: What They Are and Where They Come From, American Spice Trade Assoc., New York, 1951, 13.
74. **Shankaracharya, N. B. and Natarajan, C. P.,** Leafy spices — chemical composition and uses, *Ind. Food Packer,* 25, 28, 1971.
75. **Winton, A. L. and Winton, K. B.,** *The Stucture and Composition of Foods,* Vol. 4, John Wiley & Sons, London, 1939.
76. **Lawrence, B. M.,** Laurel leaf oil, *Perfumer Flavorist,* 12 (Aug.—Sept.), 71—73, 1987.

77. **Movesti, G.,** Laurel (*Laurus nobilis* L.) and its utilization *Revital Escenze Profumi, Piante Officolii vegetali saponi*, 23, 220—249, 1941; *Chem. Abstr.*, 83328, 1941.
78. **Meroŕy, J.,** *Food Flavorings, Composition Manufacture and Uses,* AVI Publ., Westport, CT, 1968, 90, 1927.
79. **Ikeda, R. M., Stanley, H., Vannier, E. S. H., and Spitler, J.,** The monoterpene hydrocarbon composition of some essential oils, *J. Food Sci.*, 27, 455—458, 1962.
80. Spices: What They Are and Where They Come From, American Spice Trade Assoc., New York, 1957, 3.
81. *Thorpey's Dictionary of Applied Chemistry,* Vol. 7, 4th ed., Longmans Green Co., New York, 1956, 9979.
82. **Pertolid, M. G. and Stancher, B.,** Characterization of essential oil of laurel, *Atti Congr. Qual.*, 303, 1967; *Chem. Abstr.*, 73, 91168g, 1970.
83. **Lawrence, B. M.,** Laurel leaf oil, *Perfumer and Flavorist,* 5, 33—34, 1980.
84. **Huang, T. C., Liu, P. K., Chang, S. Y., Chou, C. G., and Tseng, H. L.,** Study of anti-asthmatic constituents in *Ocimum basilicum* Benth, *Yao Hsueh T'ung Pao,* 16, 56, 1981.
85. **Lawrence, B. M., Lawrence, R., Powell, H., and Peele, D. M.,** Variation in the Genus *Ocimum,* Paper No. 34, 8th Int. Congr. Essential Oils, Cannes, October, 1980.
86. **Shankaracharya, N. B. and Natarajan, C. P.,** Leafy spices, chemical composition and uses, *Indian Food Packer,* 25, 30, 1971.
87. **Kharebava, L. G. and Sardzhveladze, K. L.,** Study of the essential oil of the laurel by the method of capillary gas chromatography, *Subtrop. Kult.*, 2, 98—102, 1979.
88. **Yoshida, T.,** On the oil containing tissue, the essential oil contents and the chemical composition of the essential oil in laurel leaf *(Laurus nobilis), Nettai Nogyo,* 23, 6—10, 1979.
89. **Lawrence, M.,** *Chemical Evaluation of Bay Oils in Essential Oils,* Allured Publishing Corp., 1979.
90. **Stanley, W. L., Ikeda, R. M., and Cook, S.,** Hydrocarbon composition of lemon oils and their relationship to optical rotation, *Food Technol.*, 15, 381, 1961.
91. **Boelens, M. H. and Sindreu, R. J.,** The chemical composition of laurel leaf oil obtained by steam distillation and hydrodiffusion, in *Progress in Essential Oil Research,* Brunke, E. J., Ed., Walter de Gruyter, Berlin, 1986, 99—110.
92. **Anac, O.,** Essential oil contents and chemical composition of Turkish laurel leaves, *Perfumer Flavorist,* 11(5), 73—75, 1986.
93. **Skrubis, B. G.,** The drying of laurel leaves, *Perfumer and Flavorist,* 7 (Oct—Nov), 37, 1982.
94. **Budzynski, S.,** Preliminary studies of the oil of *Monarda fistulosa, Chem. Abstr.*, 28, 1815, 1934.
95. **Asslani, U.,** Essential oils from *Laurus nobilis* L. Bull, Univ. Shtelenon, Tirawes, Ser., *Shkençat Natyr.*, 23(3), 93—113, 1969.
96. **Lamparsky, D.,** Etudes sur les matieres vegetables. CLXXXII. Sur les alcohols present dans L'huille essentielle de bay, *Helv. Chim. Acta,* 46, 185—187, 1963.
97. **Mazza, G., Chubey, B. B., and Kichu,** Essential oil of *Monarda fistulosa* L. var menthaefolia a potential source of Geraniol, *Flavor Fragrance J.*, 2, 129—132, 1987.
98. **Ivanova, Yu, I.,** The chemical composition of some spring annuals of Turkmenistan, *Izvest. Akad. Naira Turkmen S.S.R.*, 1, 85—87, 1954; *Chem. Abstr.*, 50, 5101h, 1956.
99. **Plakhova, N. B.,** Comparative effects of tannins from Siberian plants on bacteria of the dysentry group, *Farmakol Toksikol* 17, 39—42, 1954; *Chem. Abstr.*, 48, 13820, 1954.
100. **Nano, G. M., Sacco, T., and Frattini, C.,** Ricerche sul genre Anthemis Primo contributo, *Essenze Deriv. Agrum.*, 93, 107—114, 1973.
101. **Nano, G. M., Sacco, T., and Frattini, C.,** Botanical and chemical research on *Anthemis nobilis* L. and some of its cultivars, Paper No. 114, Sixth Int. Essential Oil Congr., San Francisco, 1974.
102. **Gupta, K. R. and Vishwanathan, R.,** *In vitro* studies of antituberculosis substances from allium species, *Antibiot. Chemother.*, 5, 18—21, 1955; *Chem. Abstr.*, 49, 6368c, 1955.
103. **Andrews, J. C. and Virer, E. T.,** The oxalic acid contents of some common foods, *Food Res.*, 16, 306—312, 1951.
104. **Erakama, J. and Kemistitchti, S.,** Colorimetric determination of vitamin C in several leaves and roots, *Summer kemisklechti*, 19A, 21—25, 1946; *Chem. Abstr.*, 41, 5584c, 1952.
105. **Kuendig, H. and Somogyi, J. L.,** Antithiamine-active substances in plant foods, *Int. 2-Vitamin Forsch.*, 34, 135—141, 1963; *Chem. Abstr.*, 61, 6277a, 1964.
106. **Macgillivray, J. H., Perdue, J. W., and Yamaguchi, M.,** Food value of some minor California vegetable crops, *Univ. Calif. Agric. Exp. Stn. Truck Crops Mimeo,* 53, 11, 1952; *Chem. Abstr.*, 46, 5742, 1952.
107. **Ranganathan, S.,** Further studies on the effect of storage on the vitamin C potency of foodstuffs, *Indian J. Med. Res.*, 23, 755—762, 1935.
108. **Sarkar, B. C. and Chavan, U. P. S.,** Strontium in some Indian vegetables, *Current Sci.*, (India), 324, 418—419, 1963.
109. **Shakaracharya, N. B. and Natarajan, C. P.,** Coriander chemistry, technology and uses; *Indian Spices,* 8, 4—13, 1971.

110. **Makarova, G. M. and Borisyuk Yu, G.,** Corriander essential oil, *Format Sevt.* Za (Kiev), 14, 43—46, 1959; *Chem. Abstr.*, 58, 3, 2320.
111. **Guenther, E.,** *The Essential Oil,* Vol. 4, D. Van Nostrand, New York, 1950, 602—614.
112. **Chakravorthy, H. L. and Chakraborthy, D. P.,** Spices of India, *Indian Agriculturist,* 8, 124—177, 1964.
113. **Dutt, S.,** The Indian curry leaf tree (*Murraya Koenigii spreng*) and its essential oil, *Indian Soap J.*, 23, 201—206, 1958.
114. **Mehrotra, R. N. and Gupta, G. N.,** Chemical examination of the essential oil from the leaves of *Murraya exotica, Indian Perfumer,* 2, 71, 1958.
115. **Nigam, S. S. and Purohit, R. M.,** Chromatography of the essential oil of *Murraya Koenigii* L., *Perfumery Essential Oil Rec.,* 52, 643, 1961.
116. **Nigam, S. S. and Purohit, R. M.,** Chemical examination of the essential oil derived from the leaves of *Murraya koenigii* Spring (Indian curry leaf), *Perfumery Essential Oil Rec.,* 52, 152—155, 1961.
117. **Prakash, V. and Natarajan, C. P.,** Studies on curry leaf (*Murraya koenigii* L.), Proc. Symp. Development and Prospects of Spice Industry in India, Association of Food Scientists and Technologists, Mysore, 1974, 65.
118. **Prakash, V. and Natarajan, C. P.,** Studies on curry leaf (*Murraya koenigii* L.), *J. Food Sci. Technol.,* 11, 284—286, 1974.
119. **Khanna, R. K., Sharma, O. S., Raina, R. M., Sinha, S., and Singh, A.,** The essential oil of *Origanum majorana* raised in saline alkali soils, *Indian Perfumer,* 29, 171—176, 1985.
120. **Lawrence, B. M.,** Marjoram oils, *Perf. Flavorist,* 6 (Oct.—Nov.), 28—32, 1981.
121. **Nicoletti, R. and Baiocchi, L.,** Sul constituenti (ossigenati) del oliodi *Origano maggiorana*; ritrovamento del cis sabinene i drato, *Ann. Chim.*, 51, 1265—1271, 1961.
122. **Dayal, B. and Purohit, R. M.,** Chemical examination of the essential oil from the seeds of *Marjorana hortensis,* Moench, *Flavor Ind.*, 2, 477—480, 1971.
123. **Vashist, V. N., Nigam, M. C., Hanada, K. L., and Gupta, G. N.,** Das atherische ol nvon *Marjorana hortensis, Riechstoffe Aromen,* 13, 61—62, 1963.
124. **Loosner, G.,** Die inhaltstoffe von *Marjorana hortensis* Moench, 2. Mitteilung; cis- und transsabinenhydrat als Bestadteile des atherishche, *Ol Pharmazie,* 22, 51—54, 1967.
125. **Karawya, M. S., Balbaa, S. I., and Hifnawy, M. S. M.,** Essential oils of certain labiaceous plants of Egypt, *Am. Perf.*, 85(8), 23—28, 1970.
126. **Rey-Jolivet, J. and Boussarie, M. F.,** Differenciation de quelques huiles essentielles presentant une constitution voisine. 11. Essence de Marjolaine (*Organum Majorana* L.) et essence d'origan (*Origanum vulgare* L.), *Plantes Med. Phytother.*, 5, 199—208, 1971.
127. **Granger, R., Passet, J., and Lamy, J.,** Sur les essences diltes 'de Marjolaine', *Riv. Ital.*, 57, 446—454, 1975.
128. **Masada, Y.,** *Analyses of Essential Oils by Gas Chromatography and Mass Spectrometry,* John Wiley & Sons, New York, 1976, 85.
129. **Croteau, R.,** Site of monoterpene biosynthesis in *Marjorana hortensis* leaves, *Plant Physiol.*, 59, 519—520, 1977.
130. **Vernon, F., Richard, H., and Sandret, F.,** Huile essentielle de marjolaine (*Marjorana hortensis* Moench) en provenance d'Egypte, *Perfums Cosmet. Aromes,* 21, 85—88, 1978.
131. **Rhyu, H. Y.,** Gas Chromatographic characterization of oregano and selected spices of the Labiatae family, *J. Food Sci.*, 44, 1373—1378, 1976.
132. **Oberdieck, R.,** Ein Beitrag zur kenntnis and Analytik von Majoran *(Marjorana hortensis Moench), Dtsch. Lebensmitt. Rundschau,* 77, 63—74, 1981.
133. **Brosche, T., Vostrowsky, O., Gemeinhardt, F., Asmus, U., and Knobloch, H.,** Uber die komponenten des atherischen Ols aus *Marjorana hortensis* Moench, *Z. Naturforsch.*, 36, 23—29, 1981.
134. **Lawrence, B. M.,** Marjoram oil, *Perf. Flavorist,* 8, 67, 1983; 9, 54—56, 1984.
135. **Sarer, E., Schefter, J. J. C., and Svendsen, A. B.,** Marota penes in the essential oil of *Origanum majorana, Planta Med.*, 46, 236—239, 1962.
136. **Lawrence, B. M.,** Marjoram oils, *Perf. Flavorist,* 9, (Feb.—Mar.), 55, 1984.
137. **Calzolari, C., Stanchey, C., and Marletta, G. P.,** *Origanum* oils and their investigation by gas-chromatographic and infrared spectroscopic analysis, *Analyst,* 93, 311—318, 1968.
138. **Punia, M. S., Verma, P. K., and Sharma, G. D.,** Evaluation of four species of mint for their oil yield, *Indian Perf.*, 31, 306—311, 1987.
139. **Sahu, B. N., Baruah, A. K. S., Bardoli, D. N., Mathur, R. K., and Baruah, J. N.,** *Mentha piperita* — a promising crop for Arunachal Pradesh (India), *Indian Perfumer,* 30, 355, 1986.
140. Essential oil of Association of India, All India Seminar on *Mentha arvensis* and *Mentha piperita* oils, Proceedings, *Indian Perfumer,* 15, 1—60, 1971.
141. **Baslas, R. K.,** Studies on the influence of various factors on the essential oil from the plants of *Mentha piperita, Flavor Industry,* 1, 185—189, 1971.

142. **Virmani, O. P. and Dutt, S. C.,** Essential oil of Japanese mint, *Indian Perfumer*, 14, 21—65, 1970.
143. **Singh, A., Singh, D. V., and Hussain, A.,** A promising strain of spearmint *(Mentha spicata), Indian Perfumer,* 28, 142—145, 1984.
144. **Chakraborti, K. K. and Bhattacharya, S. C.,** Chemical examination of Indian spearmint oil. Part 1, *Perfumery Essential Oil Rec.*, 45, 217—218, 1954.
145. **Dhingra, D. R., Gupta, G. N., Chandra, G., and Patwardhan, V. M.,** Chemical examination of spearmint oil, *Indian Perf.*, 1, 21, 1957.
146. **Handa, K. L., Chopra, I. C., and Sobti, S. N.,** Aromatic plant resources of Jammu and Kashmir, *J. Sci. Ind. Res.*, 16A, 1—28, 1956.
147. **Suri, R. K. and Jain, P. P.,** Studies on oil of *Mentha spicata*, grown in Dehra Dun, *Indian Perf.*, 25, 127—128, 1981.
148. **Lawrence, B. M., Hogg, J. W., and Terhume, S. J.,** Essential oils and their constituents. X. Some new trace constituents in the oil of *Mentha piperita* L., *Flavor Industry*, 3, 467—472, 1972.
149. **Maffei, M.,** Environmental factors affecting the oil composition of some mentha species grown in Northwest Italy, *Flavour Fragrance J.*, 3, 79—85, 1988.
150. **Baslas, R. K. and Baslas, K. K.,** Chemistry of Indian essential oils — part VIII, *Flavour Ind.*, 1(7), 473, 1970.
151. **Baslas, R. K. and Baslas, R. K.,** Essential oils from exotic plants raised in Kumaon, Part I, *Perfumery Essential Oil Rec.*, 59, 110—113, 1968.
152. **Baslas, K. K. and Baslas, R. K.,** Essential oils from some exotic plants raised in Kumaon (*Citronella Java* type) Part IV, *Perfumery Essential Oil Rec.*, 60, 341—344, 1969.
153. **Sharma, M. L., Sharma, O. S., and Singh, V. M.,** Menthol from indigenously grown *Mentha arvensis, Indian Perf.*, 15, 45—47, 1971.
154. **Randhawa, G. S., Mahey, R. K., Sidhu, B. S., and Saini, S. S.,** Herb and oil yields of *Mentha spicata*, under different row spacings and nitrogen levels in Punjab (India), *Indian Perf.*, 28, 146—149, 1984.
155. **Piper, J. J. and Price, M. J.,** Atypical oils from *Mentha arvensis* var. *piperascens* (Japanese mint) plants grown from seed, *Int. Flavors Food Addit.*, 6, 196—198, 1975.
156. **Sinha, G. K. and Gupta, R.,** Essential oil of *Mentha longifolia, Flavour Ind.*, 2(5), 310—311, 1971.
157. **Chopra, I. C. and Kapoor, L. D.,** Cultivation of mint, *Farm Bull. (News Series),* 1—27, 1967.
158. **Virmani, O. P. and Dutt, S. C.,** Prospects for production of essential oils of *Mentha spicata* (spearmint) at Lucknow, *Flavour Ind.*, 1, 549—550, 1971.
159. **Mital, S. P. and Singh, R. P.,** Present status of mint cultivation in India, *Indian J. Agron.*, 15(4), 325—329, 1970.
160. **Kapoor, L. D. and Chopra, I. C.,** Cultivation of Japanese mint, pamphlet, Regional Research Laboratory, Jammu and Kashmir (India), 1959, 1—11.
161. **Gupta, R.,** Japanese mint — a new industrial crop in India, *Indian Perf.*, 19 (Part 1), 7—16, 1965.
162. **Gupta, R.,** Cultivation and distillation of Japanese mint in India, *Indian Farming*, 22, 18—23, 1972.
163. Tropical Products Institute, London. The market prospects for peppermint, arvensis and spearmint oils with particular reference to the United Kingdom market, *TPI Rep.*, No. 58/61, 1—12, 1961.
164. **Gupta, B. N., Dhingra, D. R., and Gupta, G. N.,** Oil of spearmint with special reference to oils of Indian origin, *Soap Perfumery Cosmet.*, 25, 279, 1952.
165. **Virmani, O. P. and Datta, S. C.,** Oil of spearmint, *Perfumery Essential Oil Rec.*, 59, 411, 1968.
166. **Maarse, H. and Van Os, F. H. L.,** Volatile oil of *Origanum vulgare* ssp. *vulgare*. I. Qualitative composition of the oil. II. Oil content and quantitative composition of the oil, *Flavor Ind.*, 4, 477—483, 1973.
167. **Maarse, H.,** Volatile oil of *Origanum vulgare* L. ssp. *vulgare* III. Changes in composition during maturation, *Flavor Ind.*, 5, 278—281, 1974.
168. **Ravid, U. and Putievsky, E.,** Constituents of essential oils from *Marjorana syriaca, Cardiothymus capitatus* and *Satureja thymbra, Planta Med.*, 48, 248—249, 1983.
169. **Lawrence, B. M.,** The botanical and chemical aspects of oregano, *Perfumer Flav.*, 9(5), 41—51, 1984.
170. **Fleisher, A., Fleisher, Z., and Abu-Rukun, S.,** Chemo-varieties of *Coridothymus capitatus* L. Rchb. Growing in Israel, *J. Food Sci. Agric.*, 35, 495—499, 1984.
171. **Ravid, U. and Putievsky, E.,** Caravacrol and thymol chemotypes of East Mediterranean wild Labiatae herbs in, *Progress in Essential Oil Research,* Brunke, J., Ed., Walter de Gruyter, Berlin, 1986.
172. **Culzolari, C., Stancher, B., and Pertoldi Marletta, G.,** *Origanum* oils and their investigation by gas chromatographic and infrared spectroscopic analysis, *Analyst,* 93, 313—318, 1968.
173. **Masada, Y.,** *Analysis of Essential Oils by Gas Chromatography and Mass Spectrometry,* John Wiley & Sons, New York, 1976.
174. **Papageorgiou, V. P.,** GLC-MS computer analysis of the essential oil of *Thymus capitatus, Planta Med.*, Suppl., 29—33, 1980.
175. **Papageorgiou, V. P. and Argyriadou, N.,** Trace constituents in the essential oil of *Thymus capitatus, Phytochemistry,* 20, 2295—2297, 1981.

176. **Kasting, R., Anderson, J., and Von Sydow, F.,** Volatile constituents in leaves of parsley, *Phytochemistry,* 11, 277, 1972.
177. **Freeman, G. G., Whenham, R. J., Self, R., and Eagles, J.,** Volatile flavor components of parsley leaves *(Petroselinum crispum M). J. Sci. Food Agric.,* 26, 465, 1975.
178. **Sundermann, H. C.,** Occurrence of UDP-apiose in parsley, *Biochim. Biophys. Acta,* 156, 435—436, 1968.
179. **Muller, R. A.,** Rosemary and rosemary oil, *Flavor Ind.,* 2, 633, 1971.
180. **Tucker, A. and Maciarello, M. J.,** Essential oil of rosemary cultivars, *Flavor Fragrance J.,* 1, 137—142, 1986.
181. **Virmani, O. P., Gulati, B. C., and Datta, S. C.,** Present production of the important essential oils in India, Part II, *Perfumery Essential Oil Rec.,* 58, 700—703, 1967.
182. **Brieskorn, C. H. and Doembling, H. J.,** *Arch. Pharm. (Weinheim),* 300, 1042—1044, 1967.
183. **Boelens, M. H.,** The essential oil from *Rosmarinus officinalis* L. *Perf. Flavorist,* 10 (Oct.—Nov.), 21—37, 1985.
184. **Lawrence, B. M.,** Rosemary oil, *Perf. Flavour,* 10, 35—36, 1986.
185. **Rasmussen, K. E., Rasmussen, S., and Svendsen, B. A.,** Quantitative variations of some components of the foliage volatile oil of *Rosmarinus officinalis* L. in the spring, *Pharm. Weekblad.,* 107, 309—313, 1972.
186. **Paulet, M.,** The aromatic plants of the Languedoc region, *Perf. Essential Oil Rec.,* 47, 232, 1956.
187. **Schwenker, G. and Klohn, W.,** Gas-chromatographic Untersuchungen uber die Zusammensetzung von Rosmarinolen, *Arch. Pharm.,* 296, 845—854, 1963.
188. **Scheffer, J. J. C., Koedm, A., and Svendson, A. B.,** Fraktionierte Saulenchromatographische Vortennung der Monoterpenkohlenwassertroffe dtherischer Ole die Gaschromatographie, *Sci. Pharm.,* 44, 119—128, 1976.
189. **Klien, E. and Rojahn, W.,** (+)-Verbenone a newly discovered component of Spanish rosemary oil, *Dragoco Rep.,* 14(4), 75—76, 1967.
190. **Tucakov, J.,** Variations of percentage of essential oil in diverse organs of the *Salvia officinalis* L. (cultivated), *Perfumery Essential Oil Rec.,* 46, 219, 1955.
191. **William, A., Jr. and Harlan, L.,** Terpenes as taxonomic characters in *Salvia* section Andibertia, *Brittania,* 19, 153—160, 1967; *Chem. Abstr.,* 70, 26352, 1969.
192. **Ivanic, V. P. and Savin, K.,** Chromatographic examination of volatile oil from *Salvia officinalis* L., *Acta Pharm. Jugoslav.,* 27, 139, 1977.
193. **Ivanic, R., Savin, K., Robinson, F., and Milchard, M. J.,** Gas chromatographic examination of volatile oil from *Salvia officinalis* L., *Acta Pharm. Jugoslav.,* 28, 65—69, 1978.
194. **Brieskorn, C. H. and Wescamp, R.,** New results on the triterpenes of sage leaves, *Congr. Sci. Pharm.,* 439, 1959; *Chem. Abstr.,* 56, 1744, 1962.
195. **Brieskorn, C. H. and Dalforth, S.,** Mono and sesquiterpinoids in essential oils of *Salvia officinalis: Duet Apothein,* 104, 1388—1392, 1964; *Chem. Abstr.,* 62, 2663c, 1965.
196. **Burgar, M. I., Karba, D., and Kikelj, D.,** ^{13}C-NMR analysis of essential oil of Dalmatian sage, *(Salvia officinalis), Farm. Vestn. (Ljubljana),* 30, 253—261, 1979.
197. **Rhyu, H. Y.,** Gas chromatographic characterization of sages of various geographic origins, *J. Food Sci.,* 44, 758—762, 1979.
198. **Tucker, A. C., Maciarello, M. J., and Howell, J. T.,** Botanical aspects of commercial sage, *Econ. Bot.,* 34, 16—19, 1980.
199. **Kustrak, D., Pejcinovic, M., Kuftinec, J., and Blazevic, N.,** Composition of essential oil and cytological investigation of sage species from the Island of Vis, *Acta Pharm. Jugoslav.,* 36, 431—436, 1986.
200. **Marlier, M., Lognay, G., Wathelet, J. P., Severin, M., and Juarez Sevilla, N.,** Etude comparative de L'huile essentielle de lepechinia meyeni (Walp) epling et de *Salvia officinalis* L., Paper presented at Colloque sur la Chimie des Terpenes, April 24—25, Grasse, 1986.
201. *Proceedings: Progress in Terpene Chemistry,* Joulain, D., Ed., Editions Frontieres, Gif. sur Yuvette, France, 1986.
202. **Vernin, O. and Metzger, J.,** Analysis of sage oils by GC-MS data bank *Salvia officinalis* L. and *Salvia lavandulaefolia Vahl., Perf. Flav.,* 11(5), 79—84, 1986.
203. **Tucker, A.,** Botanical nomenclature of culinary herbs and potherbs, in *Herbs, Spices, and Medicinal Plants: Recent Advances in Botany, Horticulture, and Pharmacology,* Vol. 1, Cracker, L. E. and Simon, J. E., Eds., Oryx Press, 1986, 33—80.
204. **Grzunov, K., Mastelic, J., and Ruzic, N.,** Terpene alcohols in β-D-glucosides in the leaves of dalmatian *Salvia officinalis Progress in Essential Oil Research,* Brunke, E. J., Ed., Walter de Gruyter, Berlin, 1986, 276—270.
205. **Putievsky, E., Ravid, U., and Dudai, N.,** The essential oil and yield components from various plant parts of *Salvia fruticosa, J. Natl. Prod.,* 49, 1013—1017, 1986.

206. **Bayrak, A. and Akgul, A.,** Composition of essential oils from Turkish *Salvia* species, *Phytochemistry,* 26, 846—847, 1987.
207. **Harvala, G., Menounos, P., and Argyriadou, N.,** Essential oil from *Salvia triloba, Fitoterapia,* 58, 353—356, 1987.
208. **Lawrence, B. M.,** Sage oil, *Perf. Flavorist,* 13, 53—56, 1988.
209. **Lawrence, B. M.,** Sage oil in progress in essential oils, *Perf. Flavorist,* 2(6), 41, 1980.
210. **Jimenez, J., Risco, S., Ruiz, T., and Zarzuelo, A.,** Hypoglycemic activity of *Salvia lavandulifolia, Planta Med.,* 260—262, 1986.
211. **Crespo, M. E., Jiminez, J., Navarro, C., and Zarzuelo, A.,** The essential oil of *Salvia lavandulifolia* subspecies *oxydon.* A study of its vegetative cycle, *Plant Med.,* 367—369, 1986.
212. **Janicsek, M.,** Hungarian essential oils, *Reichsfoffind,* 3, 181—234, 1928; *Chem. Abstr.,* 24, 2235, 1930.
213. **Sinha, G. K., Chauhan, N. S., and Pathak, R. C.,** Antibacterial and antifungal properties of some essential oils and some of their constituents from plants of Kumaon, *Indian Perf.,* 13, 17—21, 1970.
214. **Brunke, E. J. and Hammerschmidt, F. J.,** Proceedings of the 15th International Symposium on Essential Oils, Svendson, A. B. and Shefter, J. C., Eds., July, 19—21, 1984.
215. **Jequier, D., Nicollier, G., Tobacchi, R., and Garnero, P. T.,** Constituents of essential oil of *Salvia stenophylla, Phytochemistry,* 19, 461, 1980.
216. **Embong, M. B., Hziyev, D., and Monlar, S.,** Essential oils from herb and spices grown in Alberta, Sage oil, *Salvia officinalis, Can. Inst. Food Sci. Technol. J.,* 10, 201—207, 1977.
217. **Lawrence, B. M.,** Sage oil, *Perf. Flavorist,* 2 (Dec.—Jan.), 50, 1977.
218. **Walt, J. M. and Breyer-Broundhijk, M. G.,** *The Medicinal and Poisonous Plants of Southern and Eastern Africa,* 2nd ed., Landai, E., Ed., S. Livingstone Ltd., Edinburgh, 1962, 527.
219. **Yaniv, Z., Dafni, A., and Palevitch, A.,** in *Aromatic Plants, Basic, and Applied Aspects,* Margaris, N., Koedam, A., and Vokon, D., Eds., Martinus Nijhoff, The Hague, 1982, 265.
220. **Putievsky, E. and Dafni, A.,** *The Book of Spices,* Masada Publ., Ramat-Gan, Israel (original in Hebrew), 1979.
221. **Dafni, A., Yaniv, Z., and Palevitch, D.,** *J. Ethnopharmacol.,* 10, 295, 1984.
222. **Putievsky, E. and Ravid, U.,** International Symposium on Conservation of Genetic Resources of Aromatic and Medicinal Plants, Oeiras, Portugal, 1984.
223. **Buil, P., Garnero, J., Guichard, G., and Konur, Z.,** Riv. Ital. Essenze. Profumi Piante Off., *Aromat. Syndets Saponi. Cosmet. Aerosols,* 59, 379, 1982.
224. **Lawrence, B. M.,** *Essential Oils,* Allured Publ., Wheaton, IL, 1981.
225. **Formacek, V. and Kubeczka, K. H.,** *Essential Oil Analysis by Capillary GC and Carbon-13 NMR Spectroscopy,* John Wiley & Sons, New York, 1982.
226. **Ravid, U. and Putievsky, E.,** Essential oils of Israel: wild species of labiate. Proc. 15th Int. Symp. Essential Oils, Nordwijkerhart, The Netherlands, July 19—21, 1984.
227. **Manjunatha, T. R., Rajamani, T. S., Ramegowda, D., and Ramaswamy, M. N.,** Essential oil from *Satueria hortensis* L., *Indian Perf.,* 8, 73, 1964.
228. **Brieskorn, C. H., Briner, M., Schlumprecht, C., and Eternhardt, K. H.,** Determination of ursolic acid and etheral oils in labiate of pharmaceutical and nutritional importance, *Arch. Pharm.,* 6, 285, 1952; *Chem. Abstr.,* 47, 7164, 1952.
229. **Herman, K.,** Antioxidative action of seasonings and the labiatic acid contained therein, *Z. Lebensing Undersuch. U. Forsu.,* 116, 224—228, 1962; *Chem. Abstr.,* 56, 13310, 1962.
230. **Leone, P. and Anglescu, E.,** The essence of *Satueria montana* of Italian origin, *Gazz. Chim. Ital.,* 51, 386—390, 1921.
231. **Igolen, G. and Sontag, D.,** Oil of mountain savory (*Satueria montana* L.), *Rev. Marques Parf. France,* 17, 109—111, 1939.
232. **Karawya, M. S., Balboa, S. I., and Hifnawy, M. S. N.,** Essential oils of certain Labiaceous (Labiatae) plants of Egypt, *Am. Perf. Cosmet.,* 85(8), 23—28, 1970.
233. **Pellecuer, J.,** La Sariette des montagnes *(Satureia montana* L. Labiees), Ph.D. thesis, Université de Montpellier, 1973.
234. **Lawrence, B. M.,** Savory oil, progress in essential oils, *Perf. Flavorist,* 3(6), 57—58, 1979.
235. **Pellecuer, J. and Garnero, J.,** Etude de l'huile essentile de *Satueria montana* (Labiees) en fonction de l'econlogie et de la physiologie de la plante, Paper No. 37, VIII Int. Congr. Essential Oils, Cannes, October 1980.
236. **Garnero, J., Buil, P., and Pellecuer, J.,** Etude de la composition chimique de l'huile essentielle de *Satueria montana* L. *(Labiees),* Paper No. 114, VIII Int. Congr. Essential Oils, Cannes, October, 1980.
237. **Chialva, F., Liddle, P. A. P., Ulian, F., and de Smedt, P.,** Indagine Sulla composizione dell'olio essenaiale di *Satueria* hortensis Linnacus coltivata in Piemonte e confronto con altre di divers origine, *Riv. Ital.,* 62, 297, 1980.
238. **Lawrence, B. M.,** Savory oil, *Perf. Flavorist,* 76—77, 1981.

239. **Rhyu, H. Y.,** Gas chromatographic characterisation of oregano and other selected spices of the Labiatae family, *J. Food Sci.*, 44, 1371—1378, 1979.
240. **Bellomaria, B. and Valentini, G.,** Composizione deil olio essenziale'd Satueria montona subsp. montana dell'Appennino marchiginae, Giorn, *Bot. Ital.*, 119, 81—83, 1985.
241. **Lawrence, B. M.,** Savory oil, *Perf. Flavorist,* 13, 46—48, 1988.
242. **Thieme, H. and Tarn, N. T.,** Accumulation and composition of essential oils in *Satueria hortensis* and *Satueria montana* and *Artemisia dracunculus, Pharmazie*, 27, 255—265, 1972; *Chem. Abstr.*, 77, 58851, 1977.
243. **Zarghami, N. S. and Russell, N. F.,** *Chem. Mikrobiol. Technol. Labensrn,* 2, 184—187, 1973.
244. **Vostrowsky, O., Michaelis, K., Ihm, H., Zitland, R., and Knoblock, K.,** Uber die Komponenton des Aetherischen Oels aus Estragon (Artemisia dracunculus L.), *Z. Lebensmitt, Unters Forsch.*, 173, 365—367, 1981.
245. **Brass, M., Mildau, G., and Jork, H.,** Neue Substanzen aus Etherischen Oelen Verschiedener Artemisia species 5, Mitt Elemicin Sowie Weitre Phenylpropen-Derivate, *GIT Suppl.*, 3, 35—42, 1983.
246. **Albasini, A., Bianchi, A., Melegari, M., Vampa, G., Pecorari, P., and Rinaldi, M.,** Indagini su Piante di *Artemisia dracunculus* Ls. 1. *(Estragone), Fittother.*, 54, 229—235, 1983.
247. **Bayrak, A. and Dogan, A.,** The research on the essential oil components from tarragon *(Artemisia dracunculus), DoGA Tarim Ormancilik Ser. D,* 10, 314—318, 1986.
248. **Lawrence, B. M.,** Tarragon oil, *Perf. Flavorist,* 13, 49—50, 1988.
249. **Mohan, V.,** Studies in essential oils: Raman Effect, Part III, *Indian Oil Soap J.*, 28, 191—196, 1963.
250. **Miquel, J. D., Richard, H. M. J., and Sandret, F. G. J.,** Volatile constituents of Moroccan thyme oil, *J. Agric. Food Chem.*, 24, 833—835, 1976.
251. **Mateo, C., Morera, M. P., Sanz, J., and Hernanadez, A.,** Estudio analitico de aceites esenciales procedentes de plantas Espanolas 1. Especies del genero Thymus, *Riv. Ital.*, 60, 621—627, 1978; 61, 135, 1979.
252. **Lawrence, B. M.,** The existence of intraspecific difference in specific genera in the Labiatae family, Paper No. 35, 8th Int. Congr. Essential Oils, Cannes, October 1980.
253. **Passel, J.,** Chemical differentiation in thyme — its nature and its importance for thyme oil, *Dragoco Rep.*, No. 10, 166—174, 1980.
254. **Lawrence, B. M.,** Thyme oil, *Perf. Flavorist,* 7, 39—40, 1982.
255. **Agarwal, I. and Mathela, C. S.,** Chemical composition of essential oil of *Thymus sepyllum* L., *Proc. Natl. Acad. Sci. (India),* 48A, 144—146, 1978.
256. **Mathela, C. S. and Agarwal, I.,** Composition of essential oil of *Thymus serpyllum* L., *J. Indian Chem. Soc.*, 57, 1249—1250, 1980.
257. **Lawrence, B. M.,** Thyme oil, *Perfumer and Flavorist,* 7, 39, 1982.

INDEX